In this book, the author aims to familiarize researchers and graduate students in both physics and mathematics with the application of non-associative algebras in physics.

Topics covered by the author range from algebras of observables in quantum mechanics, angular momentum and octonions, division algebra, triple-linear products and Yang–Baxter equations. The author also covers non-associative gauge theoretic reformulation of Einstein's general relativity theory, and so on. Much of the material found in this book is not available in other standard works.

The book will be of interest to graduate students and research scientists in physics and mathematics.

MONTROLL MEMORIAL LECTURE SERIES IN MATHEMATICAL PHYSICS

Elliott W. Montroll, whose contributions to science are honored in this lecture Series in Mathematical Physics, was from 1966 until his retirement in 1981 the Einstein Professor of Physics at the University of Rochester, where the Montroll Lectures are held. Montroll was also at various times Distinguished Professor in the Institute of Physical Science and Technology of the University of Maryland, Distinguished Professor of Physics of the University of California at Irvine, Director of General Sciences at the IBM Corporation, and Vice President for Research of the Institute for Defense Analyses. At Rochester he was founder of the Institute of Fundamental Studies and maintained a vigorous and varied research program.

Montroll's research career began in 1941 with the earliest publication of the diagrammatic resummation procedure that subsequently became so prevalent in theoretical physics, and the introduction of the moment trace method for evaluating the vibrational frequency spectrum of a lattice. Among his other contributions to mathemtical physics were his development of the transfer matrix method for calculating the partition function of an interacting lattice system; the exact calculation of the spin–spin correlation functions of the two-dimensional Ising lattice using the Pfaffian method; the development during the Second World War of the control system used to stabilize the cascade separation of uranium isotopes; the first exact evaluation of the vibrational frequency spectrum of a two-dimensional lattice and pioneering work on the theory of vibrations of crystals with defects; major contributions to random walk theory and its application to physical and chemical problems, for example, the exact solution to Polya's problem about the probability of return to the origin of a three-dimensional lattice. His introduction and George Weiss of the notion of a continuous-time random walk and the associated pausing time distribution led to ground-breaking advances in the theory of transport and relaxation in disordered systems.

Among his many other activities, Montroll was twice the Lorentz Professor at Leiden, the Gibbs Lecturer for the American Mathematical Society, founder and first editor of the Journal of Mathematical Physics, recipient with Robert Herman of the Lancaster Prize and a member of the US National Academy of Sciences. His influence on the development of mathematical and chemical physics was widely appreciated through his exceptionally lucid articles and lectures, blending common sense and beautiful mathematics.

The Montroll Memorial Lecture Series has been established by Elliott Montroll's friends to provide a forum for the presentation of new developments and coherent overviews in mathematical physics. The lectures are given at the University of Rochester. It is appropriate that the Lectures will be available in book form as a continuing representation of Montroll's vitality and curiosity and intellectual commitment to the understanding and explaning of the world of science.

A. Das, J. H. Eberly, M. M. Shahin, H. M. Van Horn

MONTROLL MEMORIAL LECTURE SERIES IN
MATHEMATICAL PHYSICS: 2

Introduction to Octonion and Other Non-Associative
Algebras in Physics

Introduction to Octonion and Other Non-Associative Algebras in Physics

SUSUMU OKUBO

Department of Physics and Astronomy
University of Rochester

CAMBRIDGE
UNIVERSITY PRESS

CAMBRIDGE UNIVERSITY PRESS
Cambridge, New York, Melbourne, Madrid, Cape Town, Singapore, São Paulo

Cambridge University Press
The Edinburgh Building, Cambridge CB2 2RU, UK

Published in the United States of America by Cambridge University Press, New York

www.cambridge.org
Information on this title: www.cambridge.org/9780521472159

First published 1995
This digitally printed first paperback version 2005

A catalogue record for this publication is available from the British Library

Library of Congress Cataloguing in Publication data

Okubo, S. (Susumu), 1930–
Introduction to octonion and other non-associative algebras in physics/Susumu Okubo.
p.   cm. – (Montroll memorial lecture series in mathematical physics: 2)
Includes bibliographical references and index.
ISBN 0 521 47215 6
1. Nonassociative algebras.   2. Mathematical physics.
3. Montroll, E. W.   I. Title.   II. Series.
QC20.7.N58038   1995
530.1′5224–dc20   94-38033   CIP

ISBN-13 978-0-521-47215-9 hardback
ISBN-10 0-521-47215-6 hardback

ISBN-13 978-0-521-01792-3 paperback
ISBN-10 0-521-01792-0 paperback

To Mary

Contents

Preface

This book originated in three lectures given at the Department of Physics and Astronomy, University of Rochester, as Montroll Lectures, on 25, 26 and 30 April, 1990, in honor of the late Professor Montroll. The present expanded version is, however, the result of a special graduate lecture course given by me in the spring semester of 1993. I am greatly indebted to our Department Chairman, Professor Paul Slattery, without whose constant encouragement this book would never have materialized. I am also grateful to Professors Ashok Das and Joseph Eberly for their various assistance in completing and reading the manuscript, and to Ms Judy Mack for a superb typing job. Finally, the book is dedicated to my wife Mary Okubo.

This book is primarily intended for graduate students and researchers interested in mathematical physics. Some knowledge of quantum mechanics and angular momentum algebra at first-year graduate level are required and will be sufficient for understanding most of the material presented in the book, although further acquaintance with Lie algebras in general would be useful.

S. Okubo
Rochester, New York

1

Introduction

The saying that God is the mathematician, so that, even with meager experimental support, a mathematically beautiful theory will ultimately have a greater chance of being correct, has been attributed to Dirac. Octonion algebra may surely be called a beautiful mathematical entity. Nevertheless, it has never been systematically utilized in physics in any fundamental fashion, although some attempts have been made toward this goal. However, it is still possible that non-associative algebras (other than Lie algebras) may play some essential future role in the ultimate theory, yet to be discovered. Some motivations and applications for non-associative physics will be given in the course of this book.

As a beginning, it may be instructive to recapitulate the history[1,2] of modern algebra briefly. Let x_0, x_1, y_0, and y_1 be real numbers. It is obvious that we have the identity

$$\left(x_0^2 + x_1^2\right)\left(y_0^2 + y_1^2\right) = (x_0 y_0 - x_1 y_1)^2 + (x_0 y_1 + x_1 y_0)^2 . \qquad (1.1)$$

By introducing the imaginary unit i (with $i^2 = -1$) and defining the complex numbers x and y by

$$x = x_0 + i x_1 , \; y = y_0 + i y_1 , \qquad (1.2)$$

as well as their absolute values by ($\|x\| \equiv |x|$ in the usual notation):

$$\|x\|^2 = x_0^2 + x_1^2 , \quad \|y\|^2 = y_0^2 + y_1^2 , \qquad (1.3)$$

we can rewrite Eq. (1.1) in a more compact form:

$$\|xy\|^2 = \|x\|^2 \|y\|^2 , \qquad (1.4)$$

where we note that

$$xy = (x_0 y_0 - x_1 y_1) + i (x_0 y_1 + x_1 y_0) . \qquad (1.5)$$

A similar identity, known as the Lagrange identity, holds true for four real constants x_j and y_j $(j = 0, 1, 2, 3)$:

$$(x_0^2 + x_1^2 + x_2^2 + x_3^2)(y_0^2 + y_1^2 + y_2^2 + y_3^2) = z_0^2 + z_1^2 + z_2^2 + z_3^2, \qquad (1.6)$$

where the z_j are given by

$$\begin{aligned}
z_0 &= x_0 y_0 - x_1 y_1 - x_2 y_2 - x_3 y_3\,, \\
z_1 &= x_0 y_1 + x_1 y_0 + x_2 y_3 - x_3 y_2\,, \\
z_2 &= x_0 y_2 + x_2 y_0 + x_3 y_1 - x_1 y_3\,, \\
z_3 &= x_0 y_3 + x_3 y_0 + x_1 y_2 - x_2 y_1\,.
\end{aligned} \qquad (1.7)$$

Equations (1.6) and (1.7) can be rewritten in the form of Eq. (1.4) if we generalize the notion of complex number appropriately. We first introduce three quantities i, j, and k, satisfying

$$i^2 = j^2 = k^2 = -1\,, \quad ij = -ji = k\,, \quad jk = -kj = i\,, \quad ki = -ik = j\,, \qquad (1.8)$$

with inner products

$$< i|i > = < j|j > = < k|k > = < 1|1 > = 1\,, \qquad (1.9)$$

while all other inner products such as $< 1|i >$ or $< i|j > (i \neq j)$ vanish. Then the generalizations of the complex numbers (1.2) are the so-called quaternions

$$\begin{aligned}
x &= x_0 1 + x_1 i + x_2 j + x_3 k\,, \\
y &= y_0 1 + y_1 i + y_2 j + y_3 k\,.
\end{aligned} \qquad (1.10)$$

Their product is calculated to be

$$z = xy = z_0 1 + z_1 i + z_2 j + z_3 k\,, \qquad (1.11)$$

where $z_j(j = 0, 1, 2, 3)$ are given by Eq. (1.7). We can now readily verify that identity (1.6) is indeed equivalent to Eq. (1.4), that is, $\|xy\|^2 = \|x\|^2 \|y\|^2$ where we have set

$$\|x\|^2 = < x|x > = x_0^2 + x_1^2 + x_2^2 + x_3^2\,. \qquad (1.12)$$

Note that Eq. (1.4) may be written alternatively as

$$< xy|xy > = < x|x >< y|y >\,. \qquad (1.13)$$

Moreover, we can verify the validity of the quadratic equation

$$x^2 - 2 < x|1 > x + < x|x > 1 = 0\,, \qquad (1.14)$$

as well as the associative law

$$(xy)z = x(yz). \tag{1.15}$$

Before going into further details we will briefly comment on two points. First, the discovery of quaternion algebra (which we denote by Q) by Hamilton[3] is regarded historically as the beginning of modern algebra, since the earlier work of Galois on the symmetric group had long been neglected. Second, Hamilton (foreshadowing Dirac's remark at the beginning of this chapter) thought that such a remarkable object as the quaternion must play an important role in physics. He devoted considerable time to the subject, without much success. We now know that he was ahead of his time. Quaternion algebra appears ubiquitously in quantum mechanics in the guise of Pauli's spin matrix

$$\sigma_1 = \begin{pmatrix} 0 & 1 \\ 1 & 0 \end{pmatrix}, \quad \sigma_2 = \begin{pmatrix} 0 & -i \\ i & 0 \end{pmatrix}, \quad \sigma_3 = \begin{pmatrix} 1 & 0 \\ 0 & -1 \end{pmatrix}. \tag{1.16}$$

Then, the correspondence

$$1 \leftrightarrow E = \begin{pmatrix} 1 & 0 \\ 0 & 1 \end{pmatrix}, \quad i \leftrightarrow \sqrt{-1}\,\sigma_1, \quad j \leftrightarrow \sqrt{-1}\,\sigma_2, \quad k \leftrightarrow -\sqrt{-1}\,\sigma_3, \tag{1.17}$$

leads to the realization of Eq. (1.8) in terms of 2×2 matrices. We then say that Pauli matrices are a 2×2 complex matrix realization of abstract quaternion algebra.

In order to proceed with our discussion, it is more convenient to change the notation for quaternion algebra Q as follows. Instead of 1, i, j, and k, we use the symbols

$$i = e_1, \quad j = e_2, \quad k = e_3, \quad 1 = e_0(\equiv e), \tag{1.18}$$

where $e_0 = e = 1$ is the unit element, that is, any $x \, \epsilon \, Q$ obeys

$$xe = ex = x. \tag{1.19}$$

Moreover, relations (1.8) can be rewritten as

$$e_j e_k = -\delta_{jk} e_0 + \sum_{\ell=1}^{3} \epsilon_{jk\ell} e_\ell \tag{1.20}$$

for $j, k = 1, 2, 3$, while Eq. (1.9) is equivalent to

$$< e_\mu | e_\nu > = \delta_{\mu\nu} \tag{1.21}$$

for $\mu, v = 0, 1, 2, 3$. Here, $\epsilon_{jk\ell}$ is the Levi–Civita symbol in three dimensions. For any $x \in Q$, we express

$$x = \sum_{\mu=0}^{3} x_\mu e_\mu = x_0 e_0 + x_1 e_1 + x_2 e_2 + x_3 e_3, \qquad (1.22)$$

as before. Then we may define the quaternionic analog of its complex conjugate by

$$\bar{x} = 2 < x|e > e - x = x_0 e_0 - x_1 e_1 - x_2 e_2 - x_3 e_3, \qquad (1.23)$$

and the quadratic equation (1.14) is rewritten as

$$x\bar{x} = \bar{x}x = < x|x > e_0, \qquad (1.24a)$$

with

$$\overline{xy} = \bar{y}\,\bar{x} \qquad (1.24b)$$

analogous to the usual complex conjugate relation $zz^* = z^*z = |z|^2$.

Returning now to the original subject, much effort has been spent in the attempt to discover the more complicated composition law satisfying Eq. (1.13). Soon after the discovery of quaternion algebra, Graves and Cayley discovered octonion algebra independently, using eight objects $e_\mu (\mu = 0, 1, 2, \ldots, 7)$ instead of four, as for quaternions. As before, $e_0 = e = 1$ is the unit element, while the multiplication rule for $e_j (j = 1, 2, \ldots, 7)$ is given by

(i) $e_j^2 = -e_0$ $(j = 1, 2, \ldots, 7)$. $\qquad\qquad\qquad (1.25a)$

(ii) $e_j e_k = -e_k e_j$, $j \neq k$ $(j, k = 1, 2, \ldots, 7)$. $\qquad (1.25b)$

(iii) The only non-zero products involving different e_j and $e_k (j \neq k)$ are

$$e_1 e_2 = e_3, \quad e_5 e_1 = e_6, \quad e_6 e_2 = e_4, \quad e_4 e_3 = e_5, \quad (1.25c)$$
$$e_4 e_7 = -e_1, \quad e_6 e_7 = -e_3, \quad e_5 e_7 = -e_2,$$

and their cyclic permutations.

Then, for two octonions x and y given by

$$x = \sum_{\mu=0}^{7} x_\mu e_\mu, \quad y = \sum_{\mu=0}^{7} y_\mu e_\mu, \qquad (1.26)$$

and for their product

$$z = xy \equiv \sum_{\mu=0}^{7} z_\mu e_\mu, \qquad (1.27)$$

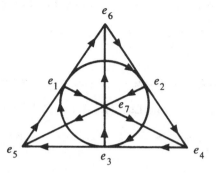

Fig. 1.1 Multiplication table for octonion.

we obtain

$$\left(\sum_{\mu=0}^{7} x_\mu^2\right)\left(\sum_{\nu=0}^{7} y_\nu^2\right) = \sum_{\mu=0}^{7} z_\mu^2. \tag{1.28}$$

Introducing the inner product by

$$< e_\mu | e_\nu > = \delta_{\mu\nu} \qquad (\mu, \nu = 0, 1, 2, \dots, 7), \tag{1.29}$$

we write Eq. (1.28) as the composition law Eq. (1.13) again, that is,

$$< x|x >< y|y > = < xy|xy >, \tag{1.30}$$

since

$$< x|x > = \sum_{\mu,\nu=0}^{7} x_\mu x_\nu < e_\mu | e_\nu > = \sum_{\mu=0}^{7} \left(x_\mu\right)^2,$$

and so on.

We can also rewrite Eqs. (1.25) as follows:

$$e_j e_k = -\delta_{jk} e_0 + \sum_{\ell=1}^{7} f_{jk\ell} e_\ell, \tag{1.31}$$

for $j, k = 1, 2, \dots, 7$, where the $f_{jk\ell}$ are totally anti-symmetric in $j, k,$ and ℓ, with values 1, 0, -1 just as is the Levi–Civita symbol $\epsilon_{jk\ell}$ for the quaternion case. Moreover, $f_{jk\ell} = +1$ for $jk\ell = 123, 516, 624, 435, 174, 376, 275$. We can depict this fact graphically[4,5] as in Fig. 1.1. Note that our definition of e_7 differs in sign from the one given by Günaydin and Gürsey.[5]

We note that octonion algebra is *not* associative, that is, the associative law Eq. (1.15) is violated for some x, y, and z. For example, we calculate

$$(e_5 e_1)e_2 = f_{516} e_6 e_2 = e_6 e_2 = f_{624} e_4 = e_4,$$

but

$$e_5(e_1 e_2) = e_5 e_3 = f_{534} e_4 = -f_{435} e_4 = -e_4 \neq (e_5 e_1)e_2.$$

Now, for a while let us assume Eqs. (1.29) and (1.31) only, that is,

$$< e_j | e_k > = \delta_{jk}, \quad < e_0 | e_0 > = 1, \quad < e_j | e_0 > = < e_0 | e_j > = 0, \quad (1.32a)$$

and

$$e_j e_k = -\delta_{jk} e_0 + \sum_{\ell=1}^{7} f_{jk\ell} e_\ell, \tag{1.32b}$$

where $f_{jk\ell}$ is totally anti-symmetric. We do *not* yet require a particular form of $f_{jk\ell}$ as in Fig. 1.1. Then, for any

$$x = x_0 e_0 + \sum_{j=1}^{7} x_j e_j,$$

$$\tag{1.33}$$

$$\bar{x} = x_0 e_0 - \sum_{j=1}^{7} x_j e_j = 2 < x|e > e - x,$$

we obtain the quadratic equation

$$x\bar{x} = \bar{x}x = < x|x > e_0, \tag{1.34a}$$

since we calculate

$$x\bar{x} = (x_0)^2 e_0 e_0 - \sum_{j,k=1}^{7} x_j x_k e_j e_k$$

$$= (x_0)^2 e_0 - \frac{1}{2} \sum_{j,k=1}^{7} x_j x_k (e_j e_k + e_k e_j),$$

but

$$e_j e_k + e_k e_j = -2\delta_{jk} e_0 \tag{1.34b}$$

because of the totally anti-symmetric condition for $f_{jk\ell}$. Hence,

$$x\bar{x} = (x_0)^2 e_0 + \sum_{j=1}^{7} (x_j)^2 e_0 = < x|x > e_0,$$

where we note that

$$< x|x > = \sum_{\mu,\nu=0}^{7} x_\mu x_\nu < e_\mu|e_\nu > = \sum_{\mu,\nu=0}^{7} x_\mu x_\nu \delta_{\mu\nu}$$

$$= \sum_{\mu=0}^{7} (x_\mu)^2 = x_0^2 + \sum_{j=1}^{7} x_j^2.$$

Next, we establish the useful relation

$$< e_\mu e_\nu|e_\lambda > = < e_\mu|e_\lambda \bar{e}_\nu > \tag{1.35}$$

for any $\mu, \nu, \lambda = 0, 1, 2, \ldots, 7$. To check this, consider first the case where neither μ, ν, nor λ is equal to zero. Then,

$$< e_j e_k|e_\ell > = < -\delta_{jk} e_0 + \sum_{m=1}^{7} f_{jkm} e_m|e_\ell >$$

$$= -\delta_{jk} < e_0|e_\ell > + \sum_{m=1}^{7} f_{jkm} < e_m|e_\ell >$$

$$= 0 + \sum_{m=1}^{7} f_{jkm}\delta_{m\ell} = f_{jk\ell},$$

while

$$< e_j|e_\ell \bar{e}_k > = < e_j| - e_\ell e_k > = < e_j|\delta_{\ell k} e_0 - \sum_{m=1}^{7} f_{\ell km} e_m >$$

$$= \delta_{\ell k} < e_j|e_0 > - \sum_{m=1}^{7} f_{\ell km} < e_j|e_m >$$

$$= 0 - \sum_{m=1}^{7} f_{\ell km}\delta_{jm} = -f_{\ell kj} = f_{jk\ell},$$

since $f_{jk\ell}$ is totally anti-symmetric in j, k, and ℓ. This proves $< e_j e_k|e_\ell >$ $= < e_j|e_\ell \bar{e}_k >$. If $e_\mu = e_j$, $e_\nu = e_k$, but $e_\lambda = e_0$, then

$$< e_j e_k|e_0 > = < -\delta_{jk} e_0 + \sum_{\ell=1}^{7} f_{jk\ell} e_\ell|e_0 >$$

$$= -\delta_{jk} < e_0|e_0 > + \sum_{\ell=1}^{7} f_{jk\ell} < e_\ell|e_0 > = -\delta_{jk},$$

while

$$< e_j|e_0 \bar{e}_k > = < e_j|\bar{e}_k > = < e_j| - e_k > = -\delta_{jk},$$

proving $< e_j e_k | e_0 > \; = \; < e_j | e_0 \bar{e}_k >$ again. For all other cases, we can verify the validity of Eq. (1.35) similarly. As a consequence of relation (1.35) we have the important relation

$$< xy|z > \; = \; < x|z\bar{y} >, \tag{1.36}$$

valid for any three x, y, z, when we express

$$x = \sum_{\mu=0}^{7} x_\mu e_\mu, \quad y = \sum_{\mu=0}^{7} y_\mu e_\mu, \quad z_\mu = \sum_{\mu=0}^{7} z_\mu e_\mu.$$

In conclusion, the quadratic equations, Eq. (1.34) and Eq. (1.36), are valid for any arbitrary totally anti-symmetric $f_{jk\ell}$.

Now, what is the most important ingredient for octonion (or quaternion) algebra? We present the following Proposition.

Proposition 1. *Suppose that we have*

(i) $x\bar{x} = \bar{x}x = \; < x|x > e_0 \quad (e_0 = \; unit \; element)$,

(ii) $< xy|z > \; = \; < x|z\bar{y} >$, \hfill (1.37)

(iii) $(xy)\bar{y} = x(y\bar{y})$.

Then, the composition law

$$< xy|xy > \; = \; < x|x >< y|y > \tag{1.38}$$

follows automatically. The converse is also correct, as we will show in the following chapter.

Proof Setting $z = xy$, we calculate

$$< xy|xy > \; = \; < xy|z > \; = \; < x|z\bar{y} >,$$

by (ii). It is also true that

$$z\bar{y} = (xy)\bar{y} = x(y\bar{y}) = x\{< y|y > e_0\} = \; < y|y > xe_0 = \; < y|y > x,$$

according to relation (iii). Therefore, we obtain

$$< xy|xy > \; = \; < x| < y|y > x > \; = \; < y|y >< x|x >,$$

where we note that $< x|\lambda x > \; = \; \lambda < x|x >$ for $\lambda = \; < y|y >$, which follows from the linearity of $< x|y >$. ∎

It is sometimes convenient to rewrite the octonionic multiplication relation, Eq. (1.31), in a more compact form, as follows. Consider four-dimensional Euclidian space with a 4 - vector a_μ and anti-symmetric tensor $f_{\mu\nu} = -f_{\nu\mu}$ ($\mu, \nu = 1, 2, 3, 4$). We assume that the tensor $f_{\mu\nu}$ is self-dual, that is, it satisfies

$$f_{\mu\nu} = {}^*f_{\mu\nu} = \frac{1}{2} \sum_{\alpha,\beta=1}^{4} \epsilon_{\mu\nu\alpha\beta} f_{\alpha\beta}, \tag{1.39}$$

where $\epsilon_{\mu\nu\alpha\beta}$ with $\epsilon_{1234} = 1$ is the totally anti-symmetric Levi–Civita symbol in four dimension. Now we impose conditions[6]

$$a_\mu a_\nu = -f_{\mu\nu} - \delta_{\mu\nu} e_0, \tag{1.40a}$$

$$f_{\mu\nu} a_\lambda = -a_\lambda f_{\mu\nu} = -\delta_{\mu\lambda} a_\nu + \delta_{\nu\lambda} a_\mu - \sum_{\alpha=1}^{4} \epsilon_{\mu\nu\lambda\alpha} a_\alpha, \tag{1.40b}$$

$$\begin{aligned} f_{\mu\nu} f_{\alpha\beta} = &-\delta_{\nu\alpha} f_{\mu\beta} + \delta_{\nu\beta} f_{\mu\alpha} - \delta_{\mu\beta} f_{\nu\alpha} + \delta_{\mu\alpha} f_{\nu\beta} \\ &- (\delta_{\mu\alpha}\delta_{\nu\beta} - \delta_{\mu\beta}\delta_{\nu\alpha} + \epsilon_{\mu\nu\alpha\beta}) e_0, \end{aligned} \tag{1.40c}$$

all of which are consistent with Eq. (1.39). Here, e_0 is the unit element. When we identify

$$\begin{aligned} e_1 = f_{23} = f_{14}, \quad e_2 = f_{31} = f_{24}, \quad e_3 = f_{12} = f_{34}, \\ e_4 = a_1, \quad e_5 = a_2, \quad e_6 = a_3, \quad e_7 = a_4, \end{aligned} \tag{1.41}$$

we can then verify that this algebra will reproduce the octonion algebra. Equations (1.40) will be used in Chapter 3 to obtain the su(2)-soliton solution, while the reason for realizing the octonion algebra in the form of Eqs. (1.40) will be explained in Chapter 6.

Remark 1.1. Both quaternions and octonions satisfy the conditions of Proposition 1. In particular, (iii) is trivial for quaternions because of the associative law.

Remark 1.2. Under the same conditions, (i), (ii), and (iii) of Proposition 1, we can also prove the validity of

$$\begin{aligned} (y\bar{y})x = y(\bar{y}x), \\ (yx)\bar{y} = y(x\bar{y}). \end{aligned} \tag{1.42}$$

Algebras satisfying the limited associative properties (iii) of Eq. (1.37) and Eq. (1.42) are called alternative.[7] Hence, both quaternion and octonion are alternative algebras. The important characteristics of these algebras are the quadratic equation (i) of Eq. (1.37) and the alternative property.

Remark 1.3. As we shall see in Chapter 3, the constants $f_{jk\ell}$ for octonion algebra may be identified with the representation matrices of Dirac–Clifford algebra in seven-dimensional space.

Remark 1.4. The $x \to \bar{x}$ satisfies (i) $\overline{xy} = \bar{y}\,\bar{x}$ and (iii) $\bar{\bar{x}} = x$. Any operation satisfying these conditions is known as involution. Also, we have $< \bar{x}|\bar{y} > = < x|y >$.

Remark 1.5. In some sense, all exceptional Lie algebras G_2, F_4, E_6, E_7, E_8 can be constructed in terms of octonions by a method known as Tits' construction.[7] Because E_8 is important for super-string theory, octonions may already be playing some role in physics, just as quarternions and Pauli spin matrices.

2

Non-associative algebras

2.1 Definition and examples

Before going into further details, it may be useful to define the notion of a non-associative algebra in a more formal way in order to avoid possible confusion.

We say that A is a non-associative algebra over a field F if the following conditions are satisfied.

(i) A is a vector space over the field F, that is, A satisfies the following:

    (a) The addition $x + y$ in A for any two x, y in A is defined and is an Abelian group operation, with its identity being the zero element. In other words, we have

$$x + 0 = 0 + x = x, \quad \text{for } x \in A,$$
$$x + y = y + x, \quad \text{for } x, y \in A,$$
$$(x + y) + z = x + (y + z) \equiv x + y + z, \quad \text{for } x, y, z \in A.$$

    (b) For any $\lambda \in F$, and $x \in A$, multiplication λx is defined in A with the usual distribution laws:

$$(\mu + \nu)x = \mu x + \nu x, \quad (\mu, \nu \in F, \ x \in A),$$
$$\mu(x + y) = \mu x + \mu y, \quad (\mu \in F, \ x, y \in A),$$

and with identification

$$0x = 0, \quad 1x = x, \quad (-1)x = -x, \quad \text{etc.}$$

(ii) A bi-linear product $x y \in A$ is defined for any two x, $y \in A$ and we have

11

(a)
$$x(y + z) = xy + xz,$$
$$(x + y)z = xz + yz,$$

for x, y, $z \in A$ and

(b)
$$(\lambda x)y = x(\lambda y) = \lambda(xy)$$

for $\lambda \in F$, and x, $y \in A$.

If we have the associative law

$$(xy)z = x(yz) \tag{2.1}$$

for any three x, y, $z \in A$ in addition to conditions (a) and (b), we say that A is an associative algebra. Otherwise, we simply call A non-associative. Note that the adjective 'non-associative' merely implies that A is not necessarily associative. When the associative law of Eq. (2.1) is violated for some elements x, y, $z \in A$, then we say that A is *not* associative.

Note

In this book, the field F is assumed to be either a real, or a complex number field, unless stated otherwise. If F is a real (or complex) field, then we say that A is a real (or complex) algebra. However, many results in this book will be valid for more general fields F.

Dimension

Since A is a vector space, we have the maximum number of linearly independent vectors (or elements) of A. This maximum number is the dimension of A (as a vector space) and is written Dim A. If Dim $A < \infty$, then A is finite dimensional. We restrict ourselves here to cases of finite-dimensional algebras, unless stated otherwise.

Examples

(1) Associative algebra. We mentioned this in Eq. (2.1).

(2) Lie algebra. Suppose that, in addition to conditions (a) and (b) for the algebra, we have the following extra conditions:

(i)
$$xy = -yx \quad \text{(anti-commutativity)},$$

(ii)
$$(xy)z + (yz)x + (zx)y = 0 \quad \text{(Jacobi identity)},$$

for any x, y, $z \in A$. Then A is called a (an abstract) Lie algebra (over the field F). It is more customary to write $[x, y]$ rather than xy in such a case.

(3) Jordan algebra. Set $x^2 \equiv xx$. If we have

    (i) $\qquad\qquad\qquad xy = yx \qquad$ (commutativity),

    (ii) $\qquad\qquad (x^2 y) x = x^2 (yx) \qquad$ (Jordan identity),

the algebra A is now called a Jordan algebra. We may use the symbol $x \cdot y$, etc., instead of xy for such a case, if we wish. In Chapter 7 we will come back to the reason why the Jordan algebra may be of some interest to physics.

2.2 Multiplication table

Assume that $\mathrm{Dim}\, A = N < \infty$ (i.e. finite dimensional). Then we can find N linearly independent elements (or vectors) $e_1, e_2, \ldots, e_N \in A$ in the vector space, which we call a basis of A, as usual. Any element x of A will then be expanded uniquely as a linear combination

$$x = \sum_{\mu=1}^{N} x_\mu e_\mu, \quad (x_\mu \in F) . \tag{2.2}$$

Since the product $e_\mu e_\nu$ is an element of A, by definition, it can also be expressed as a sum:

$$e_\mu e_\nu = \sum_{\lambda=1}^{N} c_{\mu\nu}^{\lambda} e_\lambda \tag{2.3}$$

for $c_{\mu\nu}^{\lambda} \in F(\mu, \nu, \lambda = 1, 2, \ldots, N)$. We call the relation of Eq. (2.3) the multiplication table of A with respect to a basis $e_1, e_2, \ldots, e_N$. The coefficients $c_{\mu\nu}^{\lambda}$ are called structure constants of A. Note that the multiplication table completely determines A. Let

$$x = \sum_{\mu=1}^{N} x_\mu e_\mu, \quad y = \sum_{\nu=1}^{N} y_\nu e_\nu .$$

Then we can write

$$xy = \sum_{\mu,\nu=1}^{N} x_\mu y_\nu e_\mu e_\nu = \sum_{\mu,\nu,\lambda=1}^{N} x_\mu y_\nu c_{\mu\nu}^{\lambda} e_\lambda ,$$

and, therefore, writing

$$z = xy = \sum_{\lambda=1}^{N} z_\lambda e_\lambda,$$

we find

$$z_\lambda = \sum_{\mu,\nu=1}^{N} c_{\mu\nu}^\lambda x_\mu y_\nu.$$

Example of multiplication table (quaternion algebra)

$$e_\mu = e_0, e_1, e_2, e_3, \quad \mathrm{Dim}\, A = 4,$$

$$e_0 e_j = e_j e_0 = e_j \quad (j = 1, 2, 3), \quad e_0 e_0 = e_0,$$

$$e_j e_k = -\delta_{jk} e_0 + \sum_{\ell=1}^{3} \epsilon_{jk\ell} e_\ell.$$

$e_\mu \backslash e_\nu$	e_0	e_1	e_2	e_3
e_0	e_0	e_1	e_2	e_3
e_1	e_1	$-e_0$	e_3	$-e_2$
e_2	e_2	$-e_3$	$-e_0$	e_1
e_3	e_3	e_2	$-e_1$	$-e_0$

In Chapter 6, we will also show that the familiar Clebsch–Gordan co-efficients for three spin-3 recouplings in su(2) will furnish the structure constants of an octonion algebra.

For specific algebras, the structure constants must satisfy some additional conditions.

Two examples are:

(i) Associative Algebra

$$(e_\mu e_\nu) e_\lambda = e_\mu (e_\nu e_\lambda).$$

This implies that

$$\left(\sum_{\alpha=1}^{N} c_{\mu\nu}^\alpha e_\alpha \right) e_\lambda = e_\mu \left(\sum_{\alpha=1}^{N} c_{\nu\lambda}^\alpha e_\alpha \right),$$

or

$$\sum_{\alpha=1}^{N} c_{\mu\nu}^{\alpha} \left(e_{\alpha} e_{\lambda} \right) = \sum_{\alpha=1}^{N} c_{\nu\lambda}^{\alpha} \left(e_{\mu} e_{\alpha} \right) ,$$

or

$$\sum_{\alpha=1}^{N} c_{\mu\nu}^{\alpha} \sum_{\beta=1}^{N} c_{\alpha\lambda}^{\beta} e_{\beta} = \sum_{\alpha=1}^{N} c_{\nu\lambda}^{\alpha} \sum_{\beta=1}^{N} c_{\mu\alpha}^{\beta} e_{\beta} .$$

The linear independence of $e_1, e_2, \ldots, e_N$ requires the validity of

$$\sum_{\alpha=1}^{N} c_{\mu\nu}^{\alpha} c_{\alpha\lambda}^{\beta} = \sum_{\alpha=1}^{N} c_{\nu\lambda}^{\alpha} c_{\mu\alpha}^{\beta} .$$

(ii) Lie algebra. We must have

$$e_{\mu} e_{\nu} = -e_{\nu} e_{\mu} ,$$

which implies

$$c_{\mu\nu}^{\lambda} = -c_{\nu\mu}^{\lambda} ,$$

and the Jacobi identity

$$\left(e_{\mu} e_{\nu} \right) e_{\lambda} + \left(e_{\nu} e_{\lambda} \right) e_{\mu} + \left(e_{\lambda} e_{\mu} \right) e_{\nu} = 0 ,$$

which is equivalent to

$$\sum_{\alpha=1}^{N} \left(c_{\mu\nu}^{\alpha} c_{\alpha\lambda}^{\beta} + c_{\nu\lambda}^{\alpha} c_{\alpha\mu}^{\beta} + c_{\lambda\mu}^{\alpha} c_{\alpha\nu}^{\beta} \right) = 0 .$$

2.3 Associator and flexible algebra

It is often convenient to use the associator[7] defined by

$$(x, y, z) \equiv (xy)z - x(yz), \tag{2.4}$$

which measures the deviation of the product xy in A from the associative law. Then any associative algebra A satisfies $(x, y, z) = 0$, of course, while the Jordan identity $x^2(yx) = (x^2 y)x$ may be rewritten as

$$(x^2, y, x) = 0 .$$

We also note that if an algebra A possesses the unit element e, then we have

$$(e, x, y) = (x, e, y) = (x, y, e) = 0 .$$

For example, $(e, x, y) = (ex)y - e(xy) = xy - xy = 0$, etc. Similarly, condition (iii) in Eq. (1.37), that is, $(xy)\bar{y} = x(y\bar{y})$, may be written as

$$(x, y, \bar{y}) = 0.$$

But $\bar{y} = 2(y, e)e - y$, and $(x, y, e) = 0$, so that it is equivalent to

$$(x, y, y) = 0, \quad \text{or} \quad (xy)y = x(yy).$$

Analogously, relations (1.42) become

$$(y, y, x) = 0, \quad \text{or} \quad (yy)x = y(yx),$$
$$(y, x, y) = 0, \quad \text{or} \quad (yx)y = y(xy)$$

respectively. As we will state shortly, any algebra satisfying these conditions defines an alternative algebra.

Flexible algebra

Any algebra obeying the flexible law

$$(x, y, z) = -(z, y, x) \tag{2.5}$$

is called a flexible algebra. Most of the interesting non-associative algebras are flexible. If we set $z = x$ in Eq. (2.5), we obtain

$$(x, y, x) = 0 \quad \text{or} \quad (xy)x = x(yx) \tag{2.6}$$

for any flexible algebra. Conversely, if A obeys only Eq. (2.6), then it also satisfies Eq. (2.5). Here, both conditions Eq. (2.5) and (2.6) are equivalent. We can prove this as follows. Since x in Eq. (2.6) is an arbitrary element of A, we can replace it by $x \rightarrow x + \lambda z$ for an arbitrary constant $\lambda \in F$ and $z \in A$. Then, Eq. (2.6) gives

$$(x, y, x) + \lambda\{(x, y, z) + (z, y, x)\} + \lambda^2(z, y, z) = 0,$$

which gives $\lambda\{(x, y, z) + (z, y, x)\} = 0$ since $(x, y, x) = 0$ and $(z, y, z) = 0$. Setting $\lambda = 1$, this gives Eq. (2.5). We often write xyx for the common element $(xy)x = x(yx)$ for any flexible algebra, that is, $xyx \equiv (xy)x = x(yx)$.

Examples. Any commutative or anti-commutative algebra is flexible. If $yx = \pm xy$, then $x(yx) = \pm(yx)x = \pm(\pm xy)x = (xy)x$, so that $(x, y, x) = 0$. In particular, both Lie and Jordan algebras are flexible. Any associative algebra is also, of course, flexible since $(x, y, z) = 0$ for any x, y, and z.

2.4 Power-associative algebra

We must be careful when defining powers, x^2, x^3, x^4, ... for non-associative algebra. We may define

$$x^2 = xx, \quad x^3 = x^2x = (xx)x, \quad x^4 = x^3x = ((xx)x)x, \dots,$$

etc., or, more generally, by induction:

$$x^{n+1} = x^n x \quad (n = 1, 2, 3, \dots). \tag{2.7}$$

Note that there is no guarantee that we have $(xx)x = x(xx)$ or $x^2x = xx^2$ in general, although the relation is valid for any flexible algebra. Moreover, if A is unital, that is, if A possesses the unit element e, then we may set, as usual,

$$x^0 = e \tag{2.8}$$

for the zeroth power. If an algebra A admits the power-associative law

$$x^n x^m = x^{n+m} \tag{2.9}$$

for any two positive integers n, and m, and for any $x \in A$, then A is called a power-associative algebra. As we will explain shortly, both alternative and Jordan algebras are power-associative.

It is known[7] that A is power-associative if only two special relations

$$x^2 x = xx^2, \quad x^3 x = x^2 x^2 \tag{2.10}$$

are valid. (Actually this fact holds for any field F of characteristic zero. In particular, then, the real and complex Jordan algebra is power-associative.)

For any unital power-associative algebra A, we can introduce the notion of the exponential by

$$e^x = \exp x = \sum_{n=0}^{\infty} \frac{1}{n!} x^n, \tag{2.11}$$

provided that we ignore the question of the convergence of the infinite summation. However, we can overcome the problem by restricting ourselves to only the so-called nilpotent elements which satisfy the condition

$$x^N = 0 \tag{2.12}$$

for some positive integer N. Note that N may depend upon the choice of x. Then the summation in the right-hand side of Eq. (2.11) becomes finite and is always well defined.

2.5 Alternative algebra

Suppose that we have

$$(x, y, z) = -(z, y, x) = -(y, x, z) = -(x, z, y), \qquad (2.13)$$

that is, (x, y, z) is totally anti-symmetric for exchanges of any two variables. Then A is called alternative. Setting $z = x$, it will then give

$$(x, y, x) = (x, x, y) = (y, x, x) = 0. \qquad (2.14)$$

Actually, the validity of Eq. (2.14) is equivalent to that of Eq. (2.13) by letting $x \rightarrow x + \lambda z$, just as in the flexible case. Such a method is known as a linearization or polarization method.

Any associative algebra is alternative. Also, octonion algebra is an alternative algebra if we interchange the role of x and y in (iii) of Eq. (1.37) and in Eq. (1.42). Therefore, two composition algebras, quaternion and octonion algebras, are both quadratic alternative algebras.

For any alternative algebra (which is flexible) we will denote the common element $(xy)x = x(yx)$ by xyx. The other two relations in Eq. (2.14) are rewritten as

$$(xx)y = x(xy), \quad y(xx) = (yx)x.$$

These, together with $(xx)x = x(xx)$ and $(yy)y = y(yy)$, etc., imply that any cubic expressions containing only two elements x and y satisfies the associative law. This fact has been generalized by Artin in the following theorem:[7]

Artin's theorem. *The subalgebra of an alternative algebra, which is generated by any two elements x and y is associative.*

Because of the theorem, we have, for example, $(xy)(xy) = x(yx)y = (xyx)y$, etc. Also, by choosing $x = y$, the theorem ensures that any alternative algebra is automatically power-associative.

Remark 2.1. The associative law involving more than three elements x, y, and z will not, in general, be valid. However, the following special relations, known as Moufang identities, are valid for any alternative algebras:

$$(xyx)z = x[y(xz)], \qquad (2.15a)$$
$$z(xyx) = [(zx)y]x, \qquad (2.15b)$$
$$(xy)(zx) = x(yz)x. \qquad (2.15c)$$

The Moufang identities are related to a geometry called non-Desarguean geometry, with some possible relevance to a generalization of the framework of quantum mechanics proposed by Günaydin, Piron, and Ruegg[8].

2.6 Algebras $A^{\pm}$

Let A be an algebra with the bi-linear product xy. In the same vector space A, we can introduce two new products by

$$[x, y] = xy - yx, \tag{2.16a}$$

$$x \cdot y = \frac{1}{2}(xy + yx). \tag{2.16b}$$

Then we obtain

$$[x, y] = -[y, x], \tag{2.17a}$$

$$x \cdot y = y \cdot x. \tag{2.17b}$$

In terms of the structure constants defined by $e_\mu e_\nu = \sum_{\lambda=1}^{N} c_{\mu\nu}^{\lambda} e_\lambda$, we have

$$[e_\mu, e_\nu] = \sum_{\lambda=1}^{N} \left(c_{\mu\nu}^{\lambda} - c_{\nu\mu}^{\lambda} \right) e_\lambda \equiv \sum_{\lambda=1}^{N} D_{\mu\nu}^{\lambda} e_\lambda, \tag{2.18a}$$

$$e_\mu \cdot e_\nu = \sum_{\lambda=1}^{N} \frac{1}{2} \left(c_{\mu\nu}^{\lambda} + c_{\nu\mu}^{\lambda} \right) e_\lambda \equiv \sum_{\lambda=1}^{N} E_{\mu\nu}^{\lambda} e_\lambda, \tag{2.18b}$$

that is, their structure constants $D_{\mu\nu}^{\lambda} = c_{\mu\nu}^{\lambda} - c_{\nu\mu}^{\lambda}$ and $E_{\mu\nu}^{\lambda} = \frac{1}{2}\left(c_{\mu\nu}^{\lambda} + c_{\nu\mu}^{\lambda} \right)$ are anti-symmetric and symmetric, respectively, for interchange of μ and ν. We name algebras with the products $[x, y]$ and $x \cdot y$ by the symbols A^+ and A^- respectively. Note that Dim $A^+ =$ Dim $A^- =$ Dim A, since $A^{\pm}$ and A have the same underlying vector space A.

If A^- satisfies the Jacobi identity, that is, if A^- is a Lie algebra, then A is called a Lie-admissible algebra. Similarly, if A^+ is a Jordan algebra, then A is called a Jordan-admissible algebra. In other words,

$$A = \text{Lie admissible} \leftrightarrow [[x, y], z] + [[y, z], x] + [[z, x], y] = 0,$$
$$A = \text{Jordan admissible} \leftrightarrow (x^2 \cdot y) \cdot x = x^2 \cdot (y \cdot x).$$

Note that $x^2 = xx = x \cdot x$.

If A is an associative algebra, then A^- and A^+ are, respectively, Lie and Jordan algebras, that is, A is both Lie and Jordan admissible. To demonstrate, we note the following identities (which are valid for any algebra):

$$[[x,y],z] + [[y,z],x] + [[z,x],y]$$
$$= (x,y,z) + (y,z,x) + (z,x,y) - (y,x,z) - (z,y,x) - (x,z,y) \qquad (2.19)$$

and

$$(x^2 \cdot y) \cdot x - x^2 \cdot (y \cdot x)$$
$$= \frac{1}{4} \{(x^2,y,x) + (y,x^2,x) - (x,y,x^2) - (y,x,x^2)$$
$$+ (x^2,x,y) - (x,x^2,y) + [y,(x,x,x)]\}. \qquad (2.20)$$

If A is associative, then $(x,y,z) = 0$, so that these imply the validity of both Jacobi and Jordan identities

$$[[x,y],z] + [[y,z],x] + [[z,x],y] = 0,$$
$$(x^2 \cdot y) \cdot x = x^2 \cdot (y \cdot x). \qquad (2.21)$$

Any Jordan algebra A^+ constructed from an associative algebra A is called a special Jordan algebra (see Chapter 7).

3

Hurwitz theorems and octonions

We now will sketch the proof of the so-called Hurwitz theorem with an application.

3.1 Hurwitz algebra

Since we did not define the notion of bi-linear symmetric non-degenerate form properly, it may be worth while doing it now. First,

$$< x|y > \; : \; A \otimes A \to F$$

is called bi-linear if we have

$$< \alpha x + \beta y|z > \; = \alpha < x|z > + \beta < y|z >,$$
$$< z|\alpha x + \beta y > \; = \alpha < z|x > + \beta < z|y >, \tag{3.1}$$

for any α, $\beta \; \epsilon \; F$ and for any x, y, $z \; \epsilon \; A$. It is called symmetric if we have, in addition,

$$< x|y > \; = \; < y|x > . \tag{3.2}$$

Next, suppose that any fixed element x satisfying the condition $< x|A > = 0$, that is, $< x|y > \; = 0$ for all $y \; \epsilon \; A$, is possible only for $x = 0$. Then the bi-linear form is said to be non-degenerate. Hereafter, we will consider only symmetric bi-linear non-degenerate forms unless otherwise stated. Let e_1, e_2, ..., e_N be a basis of the vector space A and set

$$< e_j|e_k > \; = g_{jk}. \tag{3.3}$$

Then, $g_{jk} = g_{kj}$ because of Eq. (3.2). Moreover, the non-degeneracy of the bi-linear form implies the existence of the inverse matrix g^{jk} satisfying

$$\sum_{\ell=1}^{N} g_{j\ell} g^{\ell k} = \delta_j^k. \tag{3.4}$$

We define an algebra A to be a Hurwitz algebra if

(i) it is unital, that is, the unit element $e \in A$ exists,

(ii) A is a composition algebra, that is, there exists a bi-linear symmetric non-degenerate form

$< x|y > \in F$ satisfying the composition law

$$< xy|xy > = < x|x >< y|y > . \qquad (3.5)$$

Then, as will be seen, the dimension of A is limited to Dim $A = 1, 2, 4,$ or 8. Moreover, the case of Dim $A = 4$ or 8 corresponds to quaternion or octonion algebra, respectively, while the two cases Dim $A = 1$ and 2 are essentially the cases of real and complex fields. This, roughly, is the content of the Hurwitz theorem.

Let us linearize Eq. (3.5) by first setting $x \to x + \lambda z$ for $\lambda \in F$ and for $z \in A$ to obtain

$$< xy|zy > = < x|z >< y|y >, \qquad (3.6)$$

where we note that $< zy|xy > = < xy|zy >$ and $< z|x > = < x|z >$. Moreover, linearizing Eq. (3.6) similarly by setting $y \to y + \lambda w$, we are led to

$$< xy|zw > + < xw|zy > = 2 < x|z >< y|w > . \qquad (3.7)$$

If we set $z = x$ and $w = y$, then Eq. (3.7) will give the original equation (3.5).

After setting $w = e$, Eq. (3.7) leads to

$$< xy|z > + < x|zy > = 2 < x|z >< y|e >,$$

which is rewritten as Eq. (1.36), that is,

$$< xy|z > = < x|z\overline{y} > . \qquad (3.8)$$

Before going into further details, we recall Eq. (1.33), where the conjugate $\overline{x}$ of x was defined by

$$\overline{x} = 2 < x|e > e - x, \qquad (3.9)$$

which is seen to satisfy

$$\overline{\overline{x}} = x \qquad (3.10)$$

when we note that $\overline{e} = e$ as well as $< e|e > = 1$. The fact that $< e|e > = 1$ follows from Eq. (3.6) with the choice of $y = e$, that is,

$< xe|ze > = < x|z >< e|e >$. Now, utilizing Eqs. (3.8) and (3.10), we rewrite

$$< xy|zw > = < xy|z\overline{\overline{w}} > = < (xy)\overline{w}|z >,$$
$$< xw|zy > = < xw|z\overline{\overline{y}} > = < (xw)\overline{y}|z >,$$

so that Eq. (3.7) becomes

$$< u|z > = 0, \qquad (3.11a)$$

where

$$u = (xy)\overline{w} + (xw)\overline{y} - 2 < y|w > x. \qquad (3.11b)$$

However, since Eq. (3.11a) holds valid for any $z \, \epsilon \, A$, the non-degeneracy of $< u|z >$ requires that $u = 0$, that is,

$$(xy)\overline{w} + (xw)\overline{y} = 2 < y|w > x. \qquad (3.12)$$

Setting $x = e$ and rewriting w as x, Eq. (3.12) gives

$$y\overline{x} + x\overline{y} = 2 < x|y > e, \qquad (3.13)$$

which can be seen to be equivalent to relation (i) of Eq. (1.37), that is, to

$$x\overline{x} = < x|x > e \qquad (3.14)$$

by further setting $y = x$ in Eq. (3.13). Using Eq. (3.9), we can also rewrite Eqs. (3.13) and (3.14) as the quadratic equations

$$xy + yx - 2 < x|e > y - 2 < y|e > x + 2 < x|y > e = 0, \qquad (3.15a)$$
$$x^2 - 2 < x|e > x + < x|x > e = 0. \qquad (3.15b)$$

From these relations, we can prove the involution property

$$\overline{xy} = \overline{y} \, \overline{x}, \qquad (3.16a)$$

when we use the fact that

$$< \overline{x}|\overline{y} > = < x|y > . \qquad (3.16b)$$

After these preparations, the choice of $w = y$ in Eq. (3.12) leads to

$$(xy)\overline{y} = < y|y > x = x(y\overline{y}). \qquad (3.17)$$

We can then also prove the validity of

$$y(\overline{y}x) = < y|y > x = (y\overline{y})x, \qquad (3.18)$$

as follows. Replacing x by $x\overline{y}$ in Eq. (3.15a), we obtain

$$(x\overline{y})y + y(x\overline{y}) = 2 < y|e > x\overline{y} + 2 < x\overline{y}|e > y - 2 < x\overline{y}|y > e.$$

Moreover, we calculate

$$y(x\bar{y}) = y\{-\bar{y}x + 2 < x|e > \bar{y} + 2 < \bar{y}|e > x - 2 < x|\bar{y} > e\},$$

as well as

$$< x\bar{y}|y > = < x|yy > = 2 < x|y >< e|y > - < y|y >< e|x > .$$

From these, we readily obtain Eq. (3.18). Equations (3.17) and (3.18) give

$$(x, y, \bar{y}) = 0 = (y, \bar{y}, x),$$

which is rewritten as

$$(x, y, y) = 0 = (y, y, x),$$

so that our algebra must be alternative. In summary, we have shown that a Hurwitz algebra is automatically a quadratic alternative algebra. Actually, the converse is also valid, as we indicated in Proposition 1 of Chapter 1.

3.2 Left and right multiplication operations

For the study of any non-associative algebra, the notion of left and right multiplication operations is very important. Let A be a non-associative algebra and let x now be a fixed element of A. Then the left multiplication L_x and right multiplication R_x are linear operators in the vector space A:

$$L_x, R_x \; : \; A \to A,$$

defined by

$$L_x y = xy, \quad R_x y = yx, \tag{3.19}$$

for all $y \in A$. We will sometimes write L_x and R_x as $L(x)$ and $R(x)$, whenever this is convenient.

It is often useful to regard L_x and R_x as matrices as follows. Let $e_0, e_1, \ldots, e_{N-1}$ be a basis of the vector space A with the multiplication table

$$e_j e_k = \sum_{\ell=0}^{N-1} C_{jk}^\ell e_\ell . \tag{3.20}$$

Setting

$$L_j \equiv L(e_j), \quad R_j \equiv R(e_j), \tag{3.21}$$

we then calculate

$$L_j e_k = e_j e_k = \sum_{\ell=0}^{N-1} C_{jk}^{\ell} e_{\ell}, \tag{3.22a}$$

$$R_j e_k = e_k e_j = \sum_{\ell=0}^{N-1} C_{kj}^{\ell} e_{\ell}. \tag{3.22b}$$

On the other hand, we recall the familiar one-to-one correspondence between a linear operator S and its matrix representation in any vector space, as follows. By expressing the effect of S as

$$S e_j = \sum_{k=0}^{N-1} e_k s_{kj}, \tag{3.23a}$$

for some constants s_{kj}, we see that its $N \times N$ matrix realization $\hat{S}$ is given by

$$< k|\hat{S}|j > = s_{kj}, \tag{3.23b}$$

so that the corresponding $N \times N$ matrices $\hat{L}_j$ and $\hat{R}_j$ are given by

$$\begin{aligned} < \ell|\hat{L}_j|k > = C_{jk}^{\ell}, \\ < \ell|\hat{R}_j|k > = C_{kj}^{\ell}. \end{aligned} \tag{3.24}$$

Since any $x \,\epsilon\, A$ can be expanded as

$$x = \sum_{j=0}^{N-1} x^j e_j,$$

for some $x^j \,\epsilon\, F$, the matrix realizations of L_x and R_x are now given by the formulae

$$< \ell|\hat{L}_x|k > = < \ell| \sum_j x^j \hat{L}_j|k > = \sum_j C_{jk}^{\ell} x^j,$$

$$< \ell|\hat{R}_x|k > = < \ell| \sum_j x^j \hat{R}_j|k > = \sum_j C_{kj}^{\ell} x^j.$$

A set consisting of all L_x and R_x ($x \,\epsilon\, A$) now forms an *associative algebra*, which is called a multiplication algebra of A. It is associative since the products of linear operators (or matrices) in a vector space are always associative. However, they are not in general commutative. For example,

$$L_x R_y z = L_x(zy) = x(zy),$$

while

$$R_y L_x z = R_y(xz) = (xz)y,$$

so that

$$[R_y, L_x]z = (xz)y - x(zy) = (x, z, y),$$

which need not be zero. Then the flexible law $(x, z, y) = -(y, z, x)$ is rewritten as

$$[R_y, L_x] = -[R_x, L_y]. \qquad (3.25)$$

Similarly, other alternative laws are rewritten, for example, as

$$(x, y, z) = -(y, x, z) \iff L_{xy} - L_x L_y = -L_{yx} + L_y L_x, \qquad (3.26a)$$

$$(x, y, z) = -(x, z, y) \iff L_{xy} - L_x L_y = L_x R_y - R_y L_x = [L_x, R_y], \qquad (3.26b)$$

$$(x, y, z) = (z, x, y) \iff L_{xy} - L_x L_y + R_{xy} - R_y R_x = 0. \qquad (3.26c)$$

Remark 3.1. In many mathematical works, including that of reference 7, the operations of linear operators to the underlying vectors are defined to be from the right to the left, in contrast to the standard convention used here. For example (instead of Eq. (3.19)), they read[7] as follows:

$$xL_y = yx, \quad xR_y = xy, \quad \text{etc.}$$

so that many relations involving L_x and/or R_x will often differ in appearance from the ones given in this book.

3.3 Dimensions of Hurwitz algebras

We will prove that the dimension of a Hurwitz algebra is restricted to 1, 2, 4, or 8. We first show the validity of

$$L_y L_{\bar{y}} = <y|y>I \qquad (3.27)$$

for any $y \in A$, where I is the identity multiplication operator defined by $Iy = y$ for any $y \in A$. This is because we have

$$L_y L_{\bar{y}} x = y(\bar{y}x) = <y|y>x = <y|y>Ix$$

by Eq. (3.18). Linearizing Eq. (3.27) by letting $y \to y + \lambda x$ gives

$$L_x L_{\bar{y}} + L_y L_{\bar{x}} = 2 <x|y>I. \qquad (3.28)$$

Now let $N = \text{Dim } A$ be the dimension of the Hurwitz algebra A. By identifying L_x and L_y with their matrix representations $\hat{L}_x$ and $\hat{L}_y$, as in the preceding section, they may be regarded as $N \times N$ matrices, while I corresponds to the $N \times N$ identity matrix.

Next, let A_0 be a sub-vector space of A defined by

$$A_0 = \{x| <x|e> = 0, \quad x \in A\}. \tag{3.29}$$

We then note that any $x \in A$ can be uniquely rewritten as $x = \lambda e + y$ ($\lambda \in F$, $y \in A_0$), with $\lambda = <e|x>$ and $y = x - <e|x>e$, so that

$$A = Fe \oplus A_0, \quad \text{Dim } A_0 = N - 1. \tag{3.30}$$

However, A_0 is not in general a sub-algebra of A since the product xy for x, $y \in A_0$ may not be an element of A_0, since we will have $<xy|e> \neq 0$ unless $<x|y> = 0$. If $x \in A_0$, then we have $\bar{x} = -x$, so that Eq. (3.28) leads to the validity of

$$L_x L_y + L_y L_x = -2 <x|y> I \tag{3.31}$$

for any x, $y \in A_0$. For simplicity, let us first assume that the underlying field F is complex. Then we can find an N-orthonormal basis $e_0(\equiv e), e_1, e_2, \ldots, e_{N-1}$ of A satisfying the orthonormal condition

$$<e_\mu|e_\nu> = \delta_{\mu\nu} \quad (\mu, \nu = 0, 1, 2, \ldots, N-1) \tag{3.32}$$

by the following reasoning: there exists an element $b \in A_0$ such that $<b|b> \neq 0$. Suppose that this is not possible and hence $<x|x> = 0$ for all $x \in A_0$. Linearizing this by setting $x \to x + \lambda y$ for $\lambda \in F$ and $y \in A_0$, this leads to $<x|y> = 0$ for any x, $y \in A_0$, which in turn gives $x = 0$ by the non-degeneracy of the bi-linear form. This proves that there exists an element b in A_0 with $<b|b> \neq 0$ if $A_0 \neq 0$. Normalizing b suitably by setting $e_1 = \lambda b$ for some $\lambda \in F$, we can set $<e_1|e_1> = 1$. Next, we define a vector space A_1 by

$$A_1 = \{x| <x|e_1> = 0, \quad x \in A_0\},$$

with $A_0 = Fe_1 \oplus A_1$, and repeat the same argument to show the existence of $e_2 \in A_1$ satisfying $<e_2|e_2> = 1$ provided that $A_1 \neq 0$. Then Eq. (3.32) is valid for $\mu, \nu = 0, 1, 2$. Repeating the same procedure, we obtain the desired result. In particular, $e_1, e_2, \ldots, e_{N-1}$ realizes an orthonormal basis of A_0. Choosing $x = e_j$ and $y = e_k$ for $j, k \neq 0$ in Eq. (3.31), we then obtain

$$L_j L_k + L_k L_j = -2\delta_{jk} I \quad (j, k = 1, 2, \ldots, N-1), \tag{3.33}$$

which is a Clifford algebra in $N - 1$-dimensional vector space A_0. We know, however, that any matrix representation of a Clifford algebra is fully reducible and, moreover, as shown by Case[9], that the unique dimension d of the irreducible representations in $N - 1$-dimensional space is given by the formula (see also Section 5.2)

$$d = 2^n$$

for $N - 1 = 2n$ (= even) and $N - 1 = 2n + 1$ (=odd). Suppose that the $N \times N$ matrix representation of L_js contains p irreducible components with its total matrix dimension of $pd = 2^n p$. This must be equal to N, so that we must have

$$N = 2^n p, \tag{3.34}$$

where n is given by either $N - 1 = 2n$ or $N - 1 = 2n + 1$. The solutions of Eq. (3.34) are found to be possible only for four cases:

$$
\begin{align}
&\text{(i)} \quad N = 1, \quad n = 0, \quad p = 1, \\
&\text{(ii)} \quad N = 2, \quad n = 0, \quad p = 2, \\
&\text{(iii)} \quad N = 4, \quad n = 1, \quad p = 2, \\
&\text{(iv)} \quad N = 8, \quad n = 3, \quad p = 1,
\end{align}
\tag{3.35}
$$

proving the desired result. Note that there is no solution of Eq. (3.34) for $N \geq 9$, since we will then have $N < 2^n$. The present proof has been adapted from that given in reference 10.

So far, we have assumed that F is a complex field. However, the same conclusion is also valid for the case where F is a real field. In that case, we extend the real field F into the complex field F_C, so that the original real Hurwitz algebra becomes a complex Hurwitz algebra without changing its dimensionality. Therefore, the dimension N of the original real Hurwitz algebra still remains equal to 1, 2, 4, or 8. Such a method is known as field extension. (However, the converse procedure is not true. The dimension of the complex algebra regarded as a real algebra is twice that of the original algebra.)

We have yet to show that the two cases $N = 4$ and $N = 8$ lead to quaternion and octonion algebras. It is not difficult to show it for the case where $N = 4$. For the general case, including that for $N = 8$, the structure constants C_{jk}^{ℓ} of the Hurwitz algebra can in principle be calculated by Eq. (3.24) from a matrix realization of the Clifford algebra.

Remark 3.2. Actually, the present result will hold for any field F. See also references 11 and 12 for the relationship between octonion and Clifford algebras.

Remark 3.3. The (N–1)-dimensional Clifford algebra is known to be intimately related to the SO(N– 1) group. Hence, octonions are somehow connected with the SO(7) group. We will come back to this point later in Chapters 6 and 8.

Remark 3.4. The multiplication tables for quaternions and octonions are rewritten more explicitly as

$$e_j e_k = L_j e_k = \sum_{\mu=0}^{N-1} e_\mu < \mu |L_j|k >$$

$$= < 0|L_j|k > e_0 + \sum_{\ell=1}^{N-1} e_\ell < \ell |L_j|k > .$$

We can choose L_j to be an anti-symmetric matrix so that

$$(L_j)^T = -L_j \quad \text{or} \quad < \ell |L_j|k > = - < k|L_j|\ell > = f_{jk\ell} ,$$

which gives $f_{jk\ell} = -f_{j\ell k}$. Also, $e_j e_k + e_k e_j = -2\delta_{jk} e_0$ gives $f_{jk\ell} = -f_{kj\ell}$. Moreover, since $e_j e_0 = e_j$, we can set

$$< 0|L_j|k > = - < k|L_j|0 > = -\delta_{jk}$$

for $j,k = 1,2,\ldots,N-1$. Therefore, we can rewrite the multiplication table as

$$e_j e_k = -\delta_{jk} e_0 + \sum_{\ell=1}^{N-1} f_{jk\ell} e_\ell ,$$

with $f_{jk\ell}$ being totally anti-symmetric in j, k, and ℓ, in accordance with the result of section 1.

Remark 3.5. Now we will briefly explain why any finite composition algebra must have only dimensions 1, 2, 4, or 8, without assuming the existence of the unit element, following the argument of Jacobson.[13] Let A be a composition (finite-dimensional) algebra which may not necessarily possess the unit element. Since $< x|y >$ is non-degenerate, there exists an element $a \in A$ such that $< a|a > \neq 0$. Then, L_a and R_a must possess their inverses $(L_a)^{-1}$ and $(R_a)^{-1}$ for the following reason. Suppose that $(L_a)^{-1}$ does not exist. Then, there exists a non-zero element $b \in A$ such that $L_a b = 0$, that is, $ab = 0$. For any $x \in A$, we calculate $0 = < ax|ab > = < a|a >< x|b >$, which gives $< x|b > = 0$. However, the non-degeneracy of the form implies that $b = 0$, which is a contradiction.

Now, we introduce a new product $x * y$ in the same vector space A by

$$x * y = (R_a^{-1}x)(L_a^{-1}y).$$

Then, $e = aa$ is the unit element of the new algebra A^*, that is, $e * x = x * e = x$. For example, we calculate

$$x * e = (R_a^{-1}x)(L_a^{-1}e).$$

But $L_a^{-1}e = L_a^{-1}(aa) = L_a^{-1}(L_a a) = a$, so that

$$x * e = (R_a^{-1}x)a = R_a(R_a^{-1}x) = x.$$

Moreover, note that

$$< L_a^{-1}x | L_a^{-1}y > = < R_a^{-1}x | R_a^{-1}y > = \frac{1}{< a|a >} < x|y >,$$

since

$$< x|y > = < L_a(L_a^{-1}x) | L_a(L_a^{-1}x) >$$
$$= < a(L_a^{-1}x) | a(L_a^{-1}y) > = < a|a >< L_a^{-1}x | L_a^{-1}y > .$$

Therefore, we obtain

$$< x * y | x * y > = < (R_a^{-1}x)(L_a^{-1}y) | (R_a^{-1}x)(L_a^{-1}y) >$$
$$= < R_a^{-1}x | R_a^{-1}x >< L_a^{-1}y | L_a^{-1}y >$$
$$= \frac{1}{< e|e >} < x|x >< y|y >,$$

since $< e|e > = < aa|aa > = < a|a >< a|a >$. Introducing a new bi-linear symmetric non-degenerate form $(x|y)$ by

$$(x|y) = \frac{1}{< e|e >} < x|y >,$$

this is equivalent to the composition law

$$(x * y | x * y) = (x|x)(y|y).$$

In other words, A^* is a composition algebra with unit element, so that Dim $A^* = 1, 2, 4,$ or 8. However, since A^* is identical to that of A as a vector space, we see that Dim $A =$ Dim $A^* = 1, 2, 4,$ or 8, proving the desired result. Note that if A is a Hurwitz algebra, then A^* will also define a Hurwitz algebra which is, however, not necessarily identical to A. Note that the unit element e of A^* may differ from that of A.

We remark that any such algebra A^* constructed from A by means of L_as and/or R_as, together with the product xy of A, is called an isotone of A in a generalized sense, since the usual definition normally *does not* include the use of the inverses $(L_a)^{-1}$ and/or $(R_a)^{-1}$. At any rate, this implies that the Hurwitz algebra is thus always of a (generalized) isotone of any composition algebra. However, the converse may not necessarily be correct for eight-dimensional composition algebras. For four-dimensional cases, it is known that any composition algebra is an isotone of the quaternion algebra if we allow, in addition, the use of the conjugate $\bar{x}$ in the product. Note that, for a Hurwitz algebra, we have $(L_a)^{-1} = \frac{1}{<a|a>} L_{\bar{a}}$ and $(R_a)^{-1} = \frac{1}{<a|a>} R_{\bar{a}}$.

3.4 An application of octonion algebra to instanton

We will present the instanton solution of the su(2) Yang–Mills field in an octonionic form, elaborating the work of Kalashinikov, *et al.*[14]

Before going into detail, let us briefly sketch the notion of the Yang–Mills gauge field.[15] Let t_a $(a = 1, 2, \ldots, N_0)$ be a matrix representation of a simple Lie algebra L specified by the Lie multiplication table

$$[t_a, t_b] = \sum_c f_{abc} t_c. \tag{3.36}$$

Then the Yang–Mills fields $A_\mu(x)$ are the Lie-algebra-valued functions given by

$$A_\mu(x) = \sum_a A_\mu^{(a)}(x) t_a \tag{3.37}$$

where the $A_\mu^{(a)}(x)$ are some functions of the coordinate. We now introduce its field strength (or curvature) tensor $F_{\mu\nu}(x)$ by

$$F_{\mu\nu}(x) = \partial_\mu A_\nu - \partial_\nu A_\mu + [A_\mu, A_\nu], \tag{3.38}$$

which transforms covariantly under a local gauge transformation

$$A_\mu(x) \to A_\mu'(x) = U^{-1}(x) A_\mu(x) U(x) + U^{-1} \partial_\mu U$$

to

$$F_{\mu\nu}(x) \to F_{\mu\nu}'(x) = U^{-1}(x) F_{\mu\nu}(x) U(x).$$

The free Yang–Mills equation of motion is now given by

$$\sum_\nu \partial^\nu F_{\mu\nu}(x) = \sum_\nu [F_{\mu\nu}, A^\nu], \tag{3.39}$$

where the repeated Greek indices imply summations in space-time indices, and where we raise the index ν by the flat metric $\eta^{\mu\nu}$. In four-dimensional space-time, it is often convenient to introduce the dual tensor ${}^*F_{\mu\nu}$ by

$$ {}^*F_{\mu\nu} = \sum_{\alpha,\beta=1}^{4} \frac{1}{2} \, \epsilon_{\mu\nu\alpha\beta} F^{\alpha\beta} , \qquad (3.40) $$

where $\epsilon_{\mu\nu\alpha\beta}$ is the totally anti-symmetric Levi–Civita symbol. Then the so-called Bianchi identity can be expressed[15] in the form (as will be explained in Remark 3.8)

$$ \sum_{\nu=1}^{4} \partial^{\nu *}F_{\mu\nu} = \sum_{\nu=1}^{4} [{}^*F_{\mu\nu}, A^\nu] . \qquad (3.41) $$

Therefore, if $F_{\mu\nu}$ satisfies the self-dual or anti-self-dual condition

$$ {}^*F_{\mu\nu} = \pm F_{\mu\nu} , \qquad (3.42) $$

it automatically satisfies the Yang–Mills equation (3.39). Unfortunately, condition (3.42) is only possible for four-dimensional Euclidian, but not for Lorentz, space-time in view of another identity

$$ {}^{**}F_{\mu\nu} = \begin{cases} F_{\mu\nu} & \text{for Euclidian 4-space} , \\ -F_{\mu\nu} & \text{for Lorentz 4-space} . \end{cases} $$

Therefore, we have to use the Euclidian space-time $x^\mu = (x^1, x^2, x^3, x^4)$ by setting $x^4 \equiv ix^0$ with $\varepsilon^{1234} = \varepsilon_{1234} = 1$.

Now, the instanton solution is a classical (i.e. c-number) solution of the Yang–Mills field equation (3.39) in Euclidian 4-space when the original Lie algebra is su(2). Therefore, we may identify

$$ t_a = \frac{i}{2} \, \sigma_a \qquad (a = 1, 2, 3) \qquad (3.43) $$

hereafter, where σ_1, σ_2, and σ_3 are familiar 2×2 Pauli matrices. As we noted, it is unnecessary to verify the validity of Eq. (3.39) if we solve the anti-self-dual condition

$$ F_{\mu\nu} = -{}^*F_{\mu\nu} = -\frac{1}{2} \sum_{\alpha,\beta=1}^{4} \epsilon_{\mu\nu\alpha\beta} F_{\alpha\beta}(x) \qquad (3.44) $$

in Euclidian 4-space with $\epsilon_{1234} = 1$. In order to obtain solutions of Eq. (3.44), we first set

$$ \partial_\mu = \left(\frac{\partial}{\partial x^1} , \frac{\partial}{\partial x^2} , \frac{\partial}{\partial x^3} , \frac{\partial}{\partial x^4} \right) = \frac{\partial}{\partial x^\mu} , \qquad (3.45) $$

and note the realization of the octonion algebra in the form of Eqs. (1.40).

We consider a tensor product $A \otimes O$ with octonion algebra O. However, for simplicity, we omit the symbol $\otimes$ and write $\sigma_a \otimes e_j = \sigma_a e_j = e_j \sigma_a$, since this will not cause confusion. We then introduce

$$\nabla = \sum_{\mu=1}^{4} a_\mu \partial_\mu = \sum_{\mu=1}^{4} e_{\mu+3} \partial_\mu , \tag{3.46}$$

$$A = \sum_{\mu=1}^{4} A_\mu \otimes a_\mu = \sum_{\mu=1}^{4} a_\mu A_\mu = \sum_{\mu=1}^{4} e_{\mu+3} A_\mu , \tag{3.47}$$

where we have written the summation symbols for the Euclidian 4-spaces explicitly in order to avoid possible confusion. Note also that A is an su(2)-Lie-algebra-valued, as well as octonionic-valued, function. Setting

$$F = \nabla A + AA , \tag{3.48}$$

and noting Eqs. (1.40a), (1.39), and (3.38), we calculate

$$F = \sum_{\mu,\nu=1}^{4} \left(\partial_\mu A_\nu + A_\mu A_\nu \right) a_\mu a_\nu = \sum_{\mu,\nu=1}^{4} \left(\partial_\mu A_\nu + A_\mu A_\nu \right) \left(- \delta_{\mu\nu} e_0 - f_{\mu\nu} \right)$$

$$= - \sum_{\mu=1}^{4} \left(\partial_\mu A_\mu + A_\mu A_\mu \right) e_0 - \frac{1}{2} \sum_{\mu,\nu=1}^{4} F_{\mu\nu} f_{\mu\nu} ,$$

which can be rewritten as

$$F = - \sum_{\mu=1}^{4} \left(\partial_\mu A_\mu + A_\mu A_\mu \right) e_0 - \frac{1}{4} \sum_{\mu,\nu=1}^{4} \left(F_{\mu\nu} + {}^* F_{\mu\nu} \right) f_{\mu\nu} \tag{3.49}$$

when we use the fact that

$$\sum_{\mu,\nu=1}^{4} F_{\mu\nu} f_{\mu\nu} = \sum_{\mu,\nu=1}^{4} F_{\mu\nu} {}^* f_{\mu\nu} = \sum_{\mu,\nu=1}^{4} {}^* F_{\mu\nu} f_{\mu\nu} .$$

Setting, for simplicity,

$$K_0(x) = \sum_{\mu=1}^{4} \left(\partial_\mu A_\mu + A_\mu A_\mu \right) , \tag{3.50}$$

it is su(2)-valued but *not* octonionic-valued. Therefore, the condition of the anti-self-duality relation (3.44) is equivalent to

$$F(x) = -K_0(x) e_0 \tag{3.51}$$

for some su(2)-valued function $K_0(x)$, that is, $F(x)$ must contain only the octonionic unit e_0. Note that the anti-self-dual condition for $F_{\mu\nu}$ naturally fits in the framework of the octonion algebra.

Now we seek the solution of Eq. (3.51) by the assumption of

$$A(x) = (\nabla f)B = -B(\nabla f) \tag{3.52}$$

for a pure function $f = f(x)$ to be determined, where B is given by

$$B = -i\sum_{k=1}^{3} \sigma_k \otimes e_k/2 = -i\sum_{k=1}^{3} e_k\sigma_k/2. \tag{3.53}$$

Note that the last relation in Eq. (3.52) is a consequence of

$$a_\lambda f_{\mu\nu} + f_{\mu\nu}a_\lambda = 0 \qquad (\mu,\nu,\lambda = 1,2,3,4), \tag{3.54}$$

as in Eq. (1.40b), or, equivalently, of

$$e_j e_k + e_k e_j = -2\delta_{jk}e_0 \tag{3.55}$$

for $j,k = 1,2,\ldots,7$.

We first observe that

$$\nabla A = \nabla(\nabla f B) = (\nabla\nabla f)B \tag{3.56}$$

because of the alternative law $(\nabla, \nabla f, B) = 0$, which is actually a simplified way of computing

$$(\nabla\nabla f)B - \nabla(\nabla f B) = \sum_{\mu,\nu=1}^{4}(a_\mu, a_\nu, B)\partial_\mu\partial_\nu f$$

$$= \frac{1}{2}\sum_{\mu,\nu=1}^{4}\{(a_\mu, a_\nu, B) + (a_\nu, a_\mu, B)\}\partial_\mu\partial_\nu f = 0$$

in view of $(x,y,z) = -(y,x,z)$ for $x = a_\mu$ and $y = a_\nu$. Then, further, we calculate

$$\nabla\nabla f = \sum_{\mu,\nu=1}^{4} a_\mu a_\nu \partial_\mu\partial_\nu f = \frac{1}{2}\sum_{\mu,\nu=1}^{4}(a_\mu a_\nu + a_\nu a_\mu)\partial_\mu\partial_\nu f$$

$$= -\left(\sum_{\mu=1}^{4}\partial_\mu\partial_\mu f\right)e_0 = -(\square f)e_0$$

by Eq. (1.40a). Therefore, we obtain

$$\nabla A = -(\square f)e_0 B = -(\square f)B. \tag{3.57}$$

Next, we calculate

$$AA = (\nabla f B)(\nabla f B) = -(\nabla f B)(B \nabla f)$$

$$= \frac{1}{4} \sum_{j,k=1}^{3} (\nabla f e_j)(e_k \nabla f) \sigma_j \sigma_k .$$

We now use the Moufang identity, Eq. (2.15c), for $x = \nabla f$, $y = e_j$, and $z = e_k$ to obtain

$$(\nabla f e_j)(e_k \nabla f) = \nabla f(e_j e_k) \nabla f .$$

However, for $j, k = 1, 2, 3$, we have

$$e_j e_k = -\delta_{jk} e_0 + \sum_{\ell=1}^{3} \epsilon_{jk\ell} e_\ell ,$$

so that

$$AA = \frac{1}{4} \sum_{j,k=1}^{3} \nabla f \left(-\delta_{jk} e_0 + \sum_{\ell=1}^{3} \epsilon_{jk\ell} e_\ell \right) \nabla f \sigma_j \sigma_k$$

$$= -\frac{3}{4} (\nabla f)(\nabla f) + \frac{i}{2} \sum_{\ell=1}^{3} (\nabla f e_\ell \nabla f) \sigma_\ell , \qquad (3.58)$$

where we have also used

$$\sigma_j \sigma_k = \delta_{jk} \, 1 + i \sum_{\ell=1}^{3} \epsilon_{jk\ell} \sigma_\ell .$$

Next, we note that, for $\ell = 1, 2, 3,$

$$\nabla f e_\ell \nabla f = \nabla f(e_\ell \nabla f) = -\nabla f(\nabla f e_\ell)$$
$$= -(\nabla f \nabla f) e_\ell + (\nabla f, \nabla f, e_\ell) = -(\nabla f \nabla f) e_\ell$$

by Eq. (3.55) for $e_\ell \nabla f = -\nabla f e_\ell$ and of the alternative law $(\nabla f, \nabla f, e_\ell) = 0$. Therefore, we can rewrite Eq. (3.58) as

$$AA = -\frac{3}{4} \nabla f \nabla f - \frac{i}{2} (\nabla f \nabla f) \sum_{\ell=1}^{3} e_\ell \sigma_\ell$$

$$= -\frac{3}{4} \nabla f \nabla f + (\nabla f \nabla f) B .$$

Moreover,

$$\nabla f \nabla f = \sum_{\mu,\nu=1}^{4} a_\mu a_\nu \partial_\mu f \partial_\nu f$$

$$= \frac{1}{2} \sum_{\mu,\nu=1}^{4} (a_\mu a_\nu + a_\nu a_\mu) \partial_\mu f \partial_\nu f = -\left(\sum_{\mu=1}^{4} \partial_\mu f \partial_\mu f \right) e_0$$

and hence

$$AA = \frac{3}{4} \left(\sum_{\mu=1}^{4} \partial_\mu f \partial_\mu f \right) e_0 - \left(\sum_{\mu=1}^{4} \partial_\mu f \partial_\mu f \right) B.$$

Adding this to Eq. (3.57), we obtain

$$F = \nabla A + AA = -K_0(x)e_0 - \left\{ \Box f + \sum_{\mu=1}^{4} \partial_\mu f \partial_\mu f \right\} B,$$

$$K_0(x) = -\frac{3}{4} \left(\sum_{\mu=1}^{4} \partial_\mu f \partial_\mu f \right).$$

Comparing this with the desired condition, Eq. (3.51), we must have

$$\Box f + \sum_{\mu=1}^{4} (\partial_\mu f) (\partial_\mu f) = 0, \tag{3.59}$$

whose solution is the famous instanton solution

$$f(x) = \ln \left[1 + \sum_{k=1}^{N} \frac{\alpha_k}{(x - \xi_k)^2} \right]. \tag{3.60}$$

Here, $\alpha_k (k = 1, 2, \ldots, N)$ are arbitrary constants and $\xi_k (k = 1, 2, \ldots, N)$ are arbitrary Euclidian 4-vectors for any positive integer N. We remark that if we wish to make the expression for B manifestly so(4) invariant, then we must introduce 4-vectors σ_μ and $\tilde{\sigma}_\mu$ by

$$\sigma_\mu = (\sigma_1, \sigma_2, \sigma_3, -iE), \quad \tilde{\sigma}_\mu = (\sigma_1, \sigma_2, \sigma_3, iE), \tag{3.61a}$$

and introduce so(4) generators $\sigma_{\mu\nu}$ by

$$\sigma_{\mu\nu} = -\frac{i}{2} \left(\sigma_\mu \tilde{\sigma}_\nu - \sigma_\nu \tilde{\sigma}_\mu \right) \tag{3.61b}$$

for $\mu, \nu = 1, 2, 3, 4$, which is self-dual, that is, $\sigma_1 = \sigma_{23} = \sigma_{14}$, $\sigma_2 = \sigma_{31} =$

σ_{24}, and $\sigma_3 = \sigma_{12} = \sigma_{34}$. Then B can be rewritten as

$$B = -\frac{i}{8} \sum_{\mu,\nu=1}^{4} f_{\mu\nu}\sigma_{\mu\nu}. \qquad (3.62)$$

Comparing Eq. (3.47) with Eq. (3.52) and using Eqs. (3.46), (3.62), and (1.40b), we can now eliminate the octonionic variables a_λ to finally obtain

$$A_\lambda(x) = \frac{i}{8} \left\{ 2\sum_{\alpha=1}^{4} \sigma_{\lambda\alpha}\partial_\alpha f(x) + \sum_{\mu,\nu,\alpha=1}^{4} \epsilon_{\lambda\mu\nu\alpha}\sigma_{\mu\nu}\partial_\alpha f(x) \right\} \qquad (3.63)$$

for the su(2) instanton solution of the Yang–Mills equation (3.39). If we wish, we can calculate the explicit form of $A_\lambda^{(a)}(x)$ $(a = 1, 2, 3)$ from Eqs. (3.37), (3.43), and (3.63), although we will not go into detail here.

Remark 3.6. We note that $(B, B, B) = 0$ but we have

$$(B, B, a_\lambda) = (a_\lambda, B, B) = -(B, a_\lambda, B) = \frac{i}{2} \sum_{\alpha,\beta,\gamma=1}^{4} \epsilon_{\lambda\alpha\beta\gamma}\sigma_{\alpha\beta} \cdot a_\gamma \neq 0.$$

Remark 3.7. The instanton solutions[16] of Belavin *et al.* are important for both physics and mathematics. The existence of the instantons may cause[17] (see, however, reference 18) the introduction of the so-called θ-vacua with accompanying possible CP violation in QCD (Quantum Chromo Dynamics). Also, the instanton plays some significant role for classifications of differentiable topology of four-dimensional manifolds.[19]

Remark 3.8. Set $D_\mu = \partial_\mu + A_\mu$ and note that $[D_\mu, D_\nu] = F_{\mu\nu}$. Then, the Jocobi identity $[[D_\mu, D_\nu], D_\lambda] + [[D_\nu, D_\lambda], D_\mu] + [[D_\lambda, D_\mu], D_\nu] = 0$ can be rewritten in the form of Eq. (3.41) in four dimensions.

Remark 3.9. Another possible link between the Yang–Mills gauge field and octonions has been attempted in reference 6. There are many other papers which utilize octonions in physics, and fairly extensive references prior to 1980 can be found in a paper by Sorgsepp and Lõhmus.[20]

Remark 3.10. Catto and Gürsey[21] have also given an interesting application of octonion algebra to the quark model, where they explored its connection to a supersymmetric theory by Miyazawa. (The author would like to express his gratitude to Professor L. C. Biedenharn for informing him of this reference.)

3.5 Derivation Lie algebras

Let A be an algebra over a field F, which is not necessarily octonionic. A linear operator

$$D : A \to A$$

satisfies, of course, the following relations:

(i) $D\,x \in A$ for $x \in A$,   (3.64a)

(ii) $D(\alpha x + \beta y) = \alpha\,D\,x + \beta\,D\,y$,   (3.64b)

for any x, $y \in A$ and for any α, $\beta \in F$. If it also obeys the condition

$$D(xy) = (Dx)y + x(Dy),   (3.65)$$

then D is called a derivation of the algebra A. Let D_1 and D_2 be two derivations of A. Then, the commutator

$$D = [D_1, D_2] \equiv D_1 D_2 - D_2 D_1$$

is also a derivation, since we calculate

$$D_1 D_2(xy) \equiv D_1\{D_2(xy)\} = D_1\{(D_2 x)y + x(D_2 y)\}$$
$$= (D_1 D_2 x)y + (D_2 x)(D_1 y) + (D_1 x)(D_2 y) + x(D_1 D_2 y).$$

Since the product of linear operators in a vector space is associative, a set consisting of all derivations of an algebra A defines a Lie algebra which is called the derivation Lie algebra of A.

We note that both D and L_x are linear operators in the vector space A, and the derivation condition (3.65) is rewritten as

$$L_{Dx} = [D, L_x]   (3.66)$$

in terms of the left multiplication operator L_x, since we calculate

$$L_{Dx}y = (Dx)y \quad \text{and} \quad [D, L_x]y = D(xy) - x(Dy).$$

Suppose, now, that A is an alternative algebra. Then, from Eqs. (3.25) and (3.26), we can first prove the validity of identities[7]

$$2[L_x, R_y] = L_{[x,y]} - [L_x, L_y] = -R_{[x,y]} - [R_x, R_y],   (3.67a)$$
$$[R_x, [R_y, R_z]] = R_w,   (3.67b)$$
$$[L_x, [L_y, L_z]] = L_w,   (3.67c)$$

where, for simplicity, we have set

$$w = [x, [y, z]] + 2(x, y, z).   (3.67d)$$

In deriving Eqs. (3.67b) and (3.67c), we used the fact that for any associative algebra we have the identity

$$[A, [B, C]] = C(AB + BA) + (AB + BA)C - B(AC + CA) - (AC + CA)B.$$

At any rate, the linear operator defined by

$$\begin{aligned} D_{x,y} &= [L_x, L_y] + [R_x, R_y] + [L_x, R_y] \\ &= L_{[x,y]} - R_{[x,y]} - 3[L_x, R_y] \end{aligned} \tag{3.68}$$

can be shown to be a derivation of any alternative algebra, that is, $D = D_{x,y}$ satisfies Eq. (3.66): $L_{Dz} = [D, L_z]$, when we utilize Eqs. (3.67).

Now let A be an octonion algebra. It can then be shown that the derivation Lie algebra generated by $D_{x,y}$ of Eq. (3.68) is isomorphic to the exceptional Lie algebra G_2, although we will not go into detail here. Conversely, we can construct an octonion algebra from the seven-dimensional irreducible representation space of G_2, as will be explained in Chapter 6. Finally, the derivation Lie algebra of the quaternion is simply su(2).

Remark 3.11. Our expressions (3.67) and (3.68) differ in signs for some terms in comparison with those of reference 7. This is due to the different multiplication rules for L_x and R_x, as has been explained in Remark 3.1 of Section 3.2.

Remark 3.12. Any derivation D which can be expressed in terms of L_xs and R_xs as in Eq. (3.68) is called an inner derivation. Otherwise, D is said to be outer.

4

Para-Hurwitz and pseudo-octonion algebras

As we recall, a Hurwitz algebra is a composition algebra with unit element. However, there are many other composition algebras without unit element. Note that their dimensions must be still 1, 2, 4, or 8, by Remark 3.5, in Section 3.3. The notable examples are pseudo-octonion and para-Hurwitz algebras. First of all, let us define para-Hurwitz algebras.[10,22]

4.1 Para-Hurwitz algebras

Let A be a Hurwitz algebra with unit element e. In the same vector space A, we introduce another product $x * y$ by

$$x * y = \bar{x}\,\bar{y}. \tag{4.1}$$

Then, it is easy to verify that the new algebra A^* obeys the following relations:

(i) $(x * y) * x = x * (y * x) = \;<x|x> y$,   (4.2a)

(ii) $<x * y|z> \; = \; <x|y * z>$,   (4.2b)

(iii) $<x * y|x * y> \; = \; <x|x><y|y>$.   (4.2c)

In particular, relations (4.2a) and (4.2c) imply that A^* is a flexible composition algebra. However, A^* does *not* possess any unit element (unless it is one dimensional), and we call A^* a para-Hurwitz algebra. With Dim $A^* = 4$ and 8, these A^* algebras are called para-quaternion and para-octonion algebras respectively. The original unit element e of A now satisfies the relation

$$x * e = e * x = \bar{x} = 2 <x|e> e - x, \tag{4.3}$$

and we call e the para-unit of the para-Hurwitz algebra A^*.

Now let us prove these assertions. First,

$$(x * y) * x = (\overline{\overline{x}\ \overline{y}})\ \overline{x} = (\overline{y}\ \overline{x})\ \overline{x} = (yx)\overline{x} = \ <x|x>\ y,$$

and

$$x * (y * x) = \overline{x}(\overline{\overline{y}\ \overline{x}}) = \overline{x}(\overline{x}\ \overline{y}) = \overline{x}(xy) = <x|x>\ y,$$

by Eqs. (3.16), (3.17), and (3.18). This proves Eq.(4.2a). Next, we calculate

$$<\overline{x}*\overline{y}|z> = \ <xy|z> = \ <x|z\overline{y}> = \ <\overline{x}|\overline{z}\overline{y}> = \ <\overline{x}|\overline{y}\ \overline{z}> = \ <\overline{x}|\overline{y}*z>$$

using Eqs. (3.8) and (3.16b). Replacing $\overline{x}$ and $\overline{y}$ by x and y, respectively, we then obtain Eq. (4.2b). Finally, we obtain

$$<x * y|x * y> = \ <\overline{x}\ \overline{y}|\overline{x}\ \overline{y}> = \ <\overline{x}|\overline{x}><\overline{y}|\overline{y}> = \ <x|x><y|y>,$$

which is Eq. (4.2c), thus completing the proof of the three assertions.

The multiplication table of the para-octonion algebra, for example, is essentially the same as that of the octonion algebra defined in Eqs. (1.32), except for the fact that we now modify the relation $e_0 e_0 = e_0$, $e_0 e_j = e_j e_0 = e_j$ by setting

$$e_0 * e_0 = e_0, \quad e_0 * e_j = e_j * e_0 = -e_j, \tag{4.4}$$

while relation (1.31) remains the same, that is,

$$e_j * e_k = -\delta_{jk} e_0 + \sum_{\ell=1}^{7} f_{jk\ell} e_\ell \quad (j,k = 1,2,\ldots,7), \tag{4.5}$$

with the same structure constants $f_{jk\ell}$.

The fact that the para-Hurwitz algebra A^* does not possess a unit element e can be shown after some calculations, although we will not go into detail here. Also, the derivation Lie algebras of para-quaternion and para-octonion algebras are su(2), and G_2, respectively, just as for the corresponding Hurwitz algebras.

4.2 Pseudo-octonion algebra

There exists another eight-dimensional composition algebra which satisfies Eqs. (4.2a)–(4.2c), but which belongs to an entirely different class of algebra. The construction of the algebra, which we call pseudo-octonion,[22,23] is given below.

Let X be any traceless 3×3 matrix, and set

$$A = \{X|X = 3 \times 3 \text{ matrix}, \text{ Tr } X = 0\}. \tag{4.6}$$

Then A is a vector space. Now, let μ and v be two complex numbers satisfying

$$3\mu v = \mu + v = 1 \tag{4.7}$$

or

$$\mu = v^* = \frac{1}{6}(3 \pm \sqrt{3}\,i). \tag{4.8}$$

Moreover, for any X, Y, ϵA, we define a non-associative product by

$$X * Y = \mu XY + vYX - \frac{1}{3}(\mathrm{Tr}\,XY)E, \tag{4.9}$$

with

$$< X|Y > = \frac{1}{6}\mathrm{Tr}(XY), \tag{4.10}$$

where E is the unit 3×3 matrix and XY implies the ordinary matrix product. When we note that $\mathrm{Tr}(X * Y) = 0$, the vector space A defines an eight-dimensional complex algebra with respect to the product $X * Y$. We calculate

$$
\begin{aligned}
(X * Y) * X &= \mu(X * Y)X + vX(X * Y) - \frac{1}{3}(\mathrm{Tr}\,(X * Y)X)E \\
&= \mu\{\mu XY + vYX - \frac{1}{3}(\mathrm{Tr}\,XY)E\}X \\
&\quad + vX\{\mu XY + vYX - \frac{1}{3}(\mathrm{Tr}\,XY)E\} \\
&\quad - \frac{1}{3}\left[\mathrm{Tr}\,\{\mu XY + vYX - \frac{1}{3}(\mathrm{Tr}\,XY)E\}X\right]E \\
&= (\mu^2 + v^2)XYX + \mu v(YX^2 + X^2Y) - \frac{1}{3}(\mu + v)(\mathrm{Tr}\,XY)X \\
&\quad - \frac{1}{3}\{(\mu + v)\mathrm{Tr}(X^2Y) - \frac{1}{3}(\mathrm{Tr}\,XY)\mathrm{Tr}\,X\}E.
\end{aligned}
$$

But since $\mathrm{Tr}\,X = 0$, and $\mu^2 + v^2 = (\mu + v)^2 - 2\mu v = 1 - \frac{2}{3} = \frac{1}{3}$, we obtain

$$(X*Y)*X = \frac{1}{3}(X^2Y + XYX + YX^2) - \frac{1}{3}(\mathrm{Tr}\,XY)X - \frac{1}{3}\mathrm{Tr}(X^2Y)E. \tag{4.11}$$

We will demonstrate that for any X, Y ϵA, we have

$$X^2Y + XYX + YX^2 - (\mathrm{Tr}\,XY)X - \mathrm{Tr}(X^2Y)E = \frac{1}{2}(\mathrm{Tr}\,X^2)Y. \tag{4.12}$$

Then Eq. (4.11) is rewritten as

$$(X * Y) * X = \frac{1}{6}(\mathrm{Tr}\,X^2)Y = <X|X>Y. \tag{4.13}$$

Similarly, we obtain

$$X * (Y * X) = <X|X> Y, \tag{4.14}$$

which yields relation (4.2a) when we change the notation suitably.

Multiplying Y by Eq. (4.12) and taking the trace, we obtain the trace identity $2 \operatorname{Tr} (X^2 Y^2) + \operatorname{Tr} (XY)^2 = \frac{1}{2} \operatorname{Tr} X^2 \operatorname{Tr} Y^2 + (\operatorname{Tr} XY)^2$, which is equivalent to $<X * Y|X * Y> = <X|X><Y|Y>$.

Now we prove the validity of Eq. (4.12). This can be done as follows. Let λ_1, λ_2, and λ_3 be three solutions of $\det |X - \lambda E| = 0$. Then, $\operatorname{Tr} X = \lambda_1 + \lambda_2 + \lambda_3 = 0$ for $X \in A$. Moreover, the Cayley–Hamilton theorem implies the validity of

$$(X - \lambda_1 E)(X - \lambda_2 E)(X - \lambda_3 E) = 0 \tag{4.15a}$$

or

$$X^3 - (\lambda_1 + \lambda_2 + \lambda_3)X^2 + (\lambda_1\lambda_2 + \lambda_2\lambda_3 + \lambda_3\lambda_1)X - \lambda_1\lambda_2\lambda_3 E = 0. \tag{4.15b}$$

But we have seen that $\lambda_1 + \lambda_2 + \lambda_3 = 0$, so that we calculate

$$\operatorname{Tr} X^2 = \lambda_1^2 + \lambda_2^2 + \lambda_3^2 = (\lambda_1 + \lambda_2 + \lambda_3)^2 - 2(\lambda_1\lambda_2 + \lambda_2\lambda_3 + \lambda_3\lambda_1)$$
$$= -2(\lambda_1\lambda_2 + \lambda_2\lambda_3 + \lambda_2\lambda_1),$$
$$\operatorname{Tr} X^3 = \lambda_1^3 + \lambda_2^3 + \lambda_3^3 = (\lambda_1 + \lambda_2 + \lambda_3)(\lambda_1^2 + \lambda_2^2 + \lambda_3^2 - \lambda_1\lambda_2 - \lambda_2\lambda_3 - \lambda_3\lambda_1)$$
$$+ 3\lambda_1\lambda_2\lambda_3 = 3\lambda_1\lambda_2\lambda_2.$$

Therefore, Eq. (4.15b) is rewritten as

$$X^3 - \frac{1}{2}(\operatorname{Tr} X^2)X - \frac{1}{3}(\operatorname{Tr} X^3)E = 0. \tag{4.16}$$

Let $\operatorname{Tr} X = \operatorname{Tr} Y = 0$, replace X by $X + \lambda Y$ for arbitrary constant λ, and pick up the coefficient of λ. This gives Eq. (4.12). We can also verify the validity of $<X * Y|Z> = <X|Y * Z>$. However, as we will show shortly, this, as well as $<X * Y|X * Y> = <X|X><Y|Y>$ is a consequence of Eqs. (4.13) and (4.14), and vice-versa.

One interesting fact is that pseudo-octonion algebra is an example of flexible Lie-admissible algebra, discussed in Chapter 2. Indeed, from Eq. (4.9), we see that

$$[X, Y]^* = X * Y - Y * X = (\mu - \nu)(XY - YX) \tag{4.17}$$

satisfies the Jacobi identity $[[X, Y]^*, Z]^* + [[Y, Z]^*, X]^* + [[Z, X]^*, Y]^* = 0$, so that its associated Lie algebra is the su(3), just as is the Lie algebra defined by any 3×3 traceless associative matrix algebra. Similarly, the derivation Lie algebra of pseudo-octonion algebra is also precisely su(3). In contrast, both octonion and para-octonion algebras are *not* Lie-admissible, although their deriviation Lie algebra is the larger algebra G_2. We can also prove that pseudo-octonion algebra possesses neither unit nor para-unit element and, hence, that it is neither a Hurwitz, nor a para-Hurwitz algebra.

The multiplication table of pseudo-octonion algebra can readily be calculated as follows. Let λ_j $(j = 1, 2, \ldots, 8)$ be the 3×3 traceless Hermitian matrices introduced by Gell-Mann.[24] Their explicit forms are

$$\lambda_1 = \begin{pmatrix} 0 & 1 & 0 \\ 1 & 0 & 0 \\ 0 & 0 & 0 \end{pmatrix}, \quad \lambda_2 = \begin{pmatrix} 0 & -i & 0 \\ i & 0 & 0 \\ 0 & 0 & 0 \end{pmatrix}, \quad \lambda_3 = \begin{pmatrix} 1 & 0 & 0 \\ 0 & -1 & 0 \\ 0 & 0 & 0 \end{pmatrix},$$

$$\lambda_4 = \begin{pmatrix} 0 & 0 & 1 \\ 0 & 0 & 0 \\ 1 & 0 & 0 \end{pmatrix}, \quad \lambda_5 = \begin{pmatrix} 0 & 0 & -i \\ 0 & 0 & 0 \\ i & 0 & 0 \end{pmatrix}, \quad \lambda_6 = \begin{pmatrix} 0 & 0 & 0 \\ 0 & 0 & 1 \\ 0 & 1 & 0 \end{pmatrix},$$

$$\lambda_7 = \begin{pmatrix} 0 & 0 & 0 \\ 0 & 0 & -i \\ 0 & i & 0 \end{pmatrix}, \quad \lambda_8 = \begin{pmatrix} \frac{1}{\sqrt{3}} & 0 & 0 \\ 0 & \frac{1}{\sqrt{3}} & 0 \\ 0 & 0 & -\frac{2}{\sqrt{3}} \end{pmatrix}, \quad (4.18)$$

which satisfy conditions

$$\text{(i)} \quad (\lambda_j)^\dagger = \lambda_j,$$
$$\text{(ii)} \quad \text{Tr}\, \lambda_j = 0, \quad (4.19)$$
$$\text{(iii)} \quad \text{Tr}(\lambda_j \lambda_k) = 2\delta_{jk},$$

for $j, k = 1, 2, \ldots, 8$. Moreover,

$$\lambda_j \lambda_k = \frac{2}{3} \delta_{jk} E + \sum_{\ell=1}^{8} \left(d_{jk\ell} + i f_{jk\ell} \right) \lambda_\ell, \quad (4.20)$$

where E is the 3×3 unit matrix, and $d_{jk\ell}$ and $f_{jk\ell}$ are totally symmetric and totally anti-symmetric real constants whose numerical values are tabulated in reference 24. Note that

$$d_{jk\ell} = \frac{1}{4} \text{Tr}\, (\lambda_j \lambda_k + \lambda_k \lambda_j)\lambda_\ell,$$

$$f_{jk\ell} = \frac{-i}{4} \text{Tr}\, (\lambda_j \lambda_k - \lambda_k \lambda_j)\lambda_\ell.$$

We remark that the $f_{jk\ell}$, defined by Eq. (4.20), are the structure constants of the su(3) Lie algebra:

$$\left[\frac{1}{2}\lambda_j, \frac{1}{2}\lambda_k\right] = i\sum_{\ell=1}^{8} f_{jk\ell}\left(\frac{1}{2}\lambda_\ell\right). \tag{4.21}$$

Now, expand $X, Y \in A$ in terms of λ_j by

$$X = \sum_{j=1}^{8} \sqrt{3}\, x_j\lambda_j, \quad Y = \sum_{j=1}^{8} \sqrt{3}\, y_j\lambda_j \tag{4.22}$$

and set

$$Z = X * Y = \sum_{j=1}^{8} \sqrt{3}\, z_j\lambda_j. \tag{4.23}$$

Then we calculate

$$<X|Y> = \sum_{j=1}^{8} x_j y_j. \tag{4.24}$$

We note here that the x_js are real numbers if X is Hermitian, that is, $X^\dagger = X$. The composition law $< X * Y | X * Y > = < X | X >< Y | Y >$ is rewritten as

$$\left(\sum_{j=1}^{8} x_j^2\right)\left(\sum_{k=1}^{8} y_k^2\right) = \sum_{j=1}^{8} z_j^2$$

with

$$z_j = \sum_{k,\ell=1}^{8} \left(\sqrt{3}\, d_{jk\ell} \mp f_{jk\ell}\right) x_k y_\ell,$$

where the $\pm$ sign corresponds to the two signs in Eq. (4.8), and where the multiplication table of pseudo-octonion algebra with respect to the orthonormal basis vectors

$$e_j = \sqrt{3}\, \lambda_j \quad (j = 1, 2, \ldots, 8) \tag{4.25}$$

is given by

$$e_j * e_k = \sum_{\ell=1}^{8} \left(\sqrt{3}\, d_{jk\ell} \mp f_{jk\ell}\right) e_\ell, \tag{4.26a}$$

with

$$< e_j | e_k > = \delta_{jk} \quad (j, k = 1, 2, \ldots, 8). \tag{4.26b}$$

We may also introduce the left and right multiplication operators L_X and R_X by

$$L_X Y = X * Y, \quad R_X Y = Y * X. \tag{4.27}$$

Then, Eqs. (4.13) and (4.14) give

$$L_X R_X = R_X L_X = <X|X> E, \tag{4.28}$$

where E is the 8×8 identity matrix. Linearizing Eq. (4.28) by letting $X \to X + \lambda Y$, we are led to

$$L_X R_Y + L_Y R_X = R_X L_Y + R_Y L_X = 2 <X|Y> E. \tag{4.29}$$

If we introduce 16×16 matrices $\Lambda(X)$ by

$$\Lambda(X) = \begin{pmatrix} 0, & L_X \\ R_X, & 0 \end{pmatrix}, \tag{4.30}$$

we are able to rewrite Eq. (4.29) as

$$\Lambda(X)\Lambda(Y) + \Lambda(Y)\Lambda(X) = 2 <X|Y> E_0, \tag{4.31a}$$

where E_0 is now the 16×16 unit matrix

$$E_0 = \begin{pmatrix} E & 0 \\ 0 & E \end{pmatrix}. \tag{4.31b}$$

Setting

$$\Lambda_j = \Lambda(e_j),$$

we can rewrite Eqs. (4.31) in the form

$$\Lambda_j \Lambda_k + \Lambda_k \Lambda_j = 2\delta_{jk} E_0 \tag{4.32}$$

for $j, k = 1, 2, \ldots, 8$. We recognize this to be the 16×16 matrix realization of the Clifford algebra in eight-dimensional carrier space. The explicit matrix realization for Λ_j is given as follows. We first introduce 8×8 matrices F_j and D_j for $j = 1, 2, \ldots, 8$ by

$$\begin{aligned} <\ell|F_j|k> &= f_{\ell jk}, \\ <\ell|D_j|k> &= d_{\ell jk}. \end{aligned} \tag{4.33}$$

Note that, for $j = 1, 2, \ldots, 8$, D_j and F_j are, respectively, real symmetric and anti-symmetric 8×8 matrices. Then, the Λ_js can be expressed as

$$\Lambda_j = \begin{pmatrix} 0, & \sqrt{3}\, D_j \mp F_j \\ \sqrt{3}\, D_j \pm F_j, & 0 \end{pmatrix}. \tag{4.34}$$

For more details, see references 10 and 25.

So far, we have assumed the underlying field F to be complex. However, we can also construct a real pseudo-octonion algebra by restricting X to be a Hermitian 3×3 traceless matrix. Let A_0 be the set of such matrices:

$$A_0 \equiv \{X | X = 3 \times 3 \text{ Hermitian matrix }, \text{ Tr } X = 0\}, \qquad (4.35)$$

which is a real sub-vector space of A defined by Eq. (4.6) with the same dimension. Since $\mu = v^*$, $X * Y$ defined by Eq. (4.9) satisfies

$$(X * Y)^\dagger = X^\dagger * Y^\dagger, \qquad (4.36)$$

where $X^\dagger$ designates the Hermitian conjugate of X. Therefore, if $X^\dagger = X$ and $Y^\dagger = Y$, then $X * Y$ is also Hermitian and, hence, A_0 defines the eight-dimensional real sub-algebra of A, which we call a real pseudo-octonion algebra. Moreover, for any $X^\dagger = X$, we have

$$< X|X > = \frac{1}{6} \text{ Tr } (XX) \geq 0 \qquad (4.37)$$

and $< X|X > = 0$ if and only if $X = 0$ identically.

Let S and T be any 3×3 Hermitian traceless matrices so that they are elements of the algebra A_0. Supposing, in addition, that $S \neq 0$, we then have $< S|S > \neq 0$. The linear equations

$$S * X = T, \quad \text{and} \quad Y * S = T \qquad (4.38)$$

for some unknown X and $Y \in A_0$ then have a unique solution in A_0 given by

$$X = \frac{1}{< S|S >} \, T * S, \qquad (4.39a)$$

$$Y = \frac{1}{< S|S >} \, S * T, \qquad (4.39b)$$

by Eqs. (4.13) and (4.14). This is an example of a division algebra, which will be discussed in more detail shortly. However, the complex pseudo-octonion algebra is not a division algebra, since for non-Hermitian matrix S we could have $< S|S > = 0$ even for $S \neq 0$.

4.3 Miscellaneous theorems on composition algebras

Both para-Hurwitz and pseudo-octonion algebras satisfy relations such as Eqs. (4.2a)–(4.2c) or (4.13) and (4.14). Changing notations, we rewrite

$x * y$ and $X * Y$, etc., simply as xy. Then Eqs. (4.2a)–(4.2c) are now rewritten in the form of

$$(xy)x = x(yx) = <x|x> y, \qquad (4.40)$$
$$<xy|z> = <x|yz>, \qquad (4.41a)$$
$$<xy|xy> = <x|x><y|y>. \qquad (4.41b)$$

First, we will prove

Theorem 1. *Let $<x|y>$ be a bi-linear symmetric non-degenerate form. Then a necessary and sufficient condition[22,23] that we have Eqs. (4.41a) and (4.41b) is the validity of Eq. (4.40).*

Proof Assume that Eqs. (4.41a) and (4.41b) hold valid. Linearizing Eq. (4.41b) by letting $y \rightarrow y + \lambda z$, we obtain

$$<xy|xz> = <x|x><y|z>.$$

Then, applying Eq. (4.41a) to this relation, we also obtain

$$<(xy)x|z> = <xy|xz> = <x|x><y|z>,$$

as well as

$$<z|x(yx)> = <zx|yx> = <z|y><x|x>.$$

Since z is arbitrary, these relations lead to the validity of Eq. (4.40) because of the non-degeneracy of the bi-linear form.

Conversely, we assume the validity of Eq. (4.40) and show that it leads to that of Eqs. (4.41a) and (4.41b) by the following reasoning. Linearizing Eq. (4.40) by letting $x \rightarrow x + \lambda z$, we obtain

$$(xy)z + (zy)x = x(yz) + z(yx) = 2 <x|z> y. \qquad (4.42)$$

We calculate the expression $(xy)(yz)$ in the following two different ways. First we set $u = yz$, so that

$$(xy)(yz) = (xy)u = 2 <x|u> y - (uy)x$$

from Eq. (4.42). However, we see that

$$uy = (yz)y = <y|y> z$$

and hence

$$(xy)(yz) = 2 <x|yz> y - <y|y> zx. \qquad (4.43)$$

Second, setting $v = xy$, we calculate

$$(xy)(yz) = v(yz) = 2 < v|z > y - z(yv)$$

and

$$yv = y(xy) = < y|y > x,$$

which leads to

$$(xy)(yz) = 2 < xy|z > y - < y|y > zx.$$

Comparing this with Eq. (4.43), we find

$$< x|yz > = < xy|z >,$$

which is precisely Eq. (4.41a). The remaining relation, Eq. (4.41b), then follows from

$$< xy|xy > = < (xy)x|y > = << x|x > y|y > = < x|x >< y|y >.$$

This completes the proof of Theorem 1. Finally, we will mention the following classification theorems. ∎

Theorem 2. *Any algebra over the field F of characteristic other than 2 satisfying the condition of Theorem 1 is either[26] a para-Hurwitz algebra or an eight-dimensional pseudo-octonion algebra.*

Theorem 3. *Any flexible algebra satisfying the composition law $< xy|xy > = < x|x >< y|y >$ for a bi-linear symmetric non-degenerate form $< x|y >$ is limited to be either[27,28] (under the same condition for F)*

   (i) *a Hurwitz algebra*

or

   (ii) *a para-Hurwitz algebra*

or

   (iii) *the eight-dimensional pseudo-octonion algebra.*

Theorem 4. *Power-associative composition algebras over the field F of characteristic other than 2 are limited only to Hurwitz algebras, that is, they must, of necessity, have the unit element.[29]*

Since proofs of these theorems are somewhat involved, they will not be given here. Interested readers are referred to references 22, as well as to 26–9.

Remark 4.1. The general classification of composition algebras, without assuming flexibility or a power-associativity law, is still unknown, although the case for $N = 4$ has been solved by Shapiro.[30]

Remark 4.2. All Hurwitz, para-Hurwitz, and pseudo-octonion algebras are examples of flexible Malcev-admissible algebras. See reference 22 for a discussion of Malcev-admissable algebras. Also, for some related topics, see papers by El-Mallah.[31]

5

Real division algebras and real Clifford algebra

5.1 Division algebras

Consider the elementary linear equation

$$ax = b$$

for any constants a and $b \in F$. We know that as long as $a \neq 0$, a solution always exists and is unique. Indeed, the solution is given by $x = a^{-1}b$.

We can generalize the concept as follows. Consider any two arbitrary but fixed elements a and b of an algebra A. In addition, we assume that $a \neq 0$. If linear equations

$$ax = b \quad \text{and} \quad ya = b$$

for two unknowns x and y always have unique solutions in A (i.e. $x, y \in A$), then the algebra A is called a division algebra over the field F. For a reason to be explained in Remark 5.2 below, the most interesting case is where F is a real field. Then we are considering real division algebras. We have already shown in the preceding chapter that the real pseudo-octonion algebra is a division algebra. Similarly, real quaternion and octonion algebras, as well as any real para-Hurwitz algebras, are division algebras, as we will demonstrate below. Consider, first, the case of real quaternion algebra. Then we know from Eq. (1.12) that

$$< x|x > = \sum_{\mu=0}^{3} (x_\mu)^2 \geq 0$$

for $x = \sum_{\mu=0}^{3} x_\mu e_\mu$ with real coefficients x_μ. Moreover, if $x \neq 0$, then $< x|x > \neq 0$ since at least one of the x_μs is non-zero. Analogously, for the real octonion algebra discussed in Chapter 1, we have

$$< x|x > = \sum_{\mu=0}^{7} (x_\mu)^2 \neq 0,$$

for any $x = \sum_{\mu=0}^{7} x_\mu e_\mu \neq 0$ with real coefficients x_μs. The proof we need of real quaternion and octonion algebras being division algebras is based upon this fact. Consider, first, the equation

$$ax = b. \tag{5.1}$$

Multiplying by $\bar{a}$ from the left, we obtain

$$\bar{a}(ax) = \bar{a}b.$$

However, for both quaternion and octonions, we know that

$$\bar{a}(ax) = (\bar{a}a)x = < a|a > x$$

from the alternative law Eq. (3.18), with $y = \bar{a}$, so that we obtain

$$< a|a > x = \bar{a}b.$$

If $a \neq 0$, we have $< a|a > \neq 0$ for real quaternion and octonion algebras and here

$$x = \frac{1}{< a|a >} \bar{a}b, \tag{5.2}$$

which implies that the solution of Eq. (5.1) is unique, if it exists. Conversely, we can easily verify that Eq. (5.2) is a solution of Eq. (5.1) since we can calculate

$$ax = \frac{1}{< a|a >} a(\bar{a}b) = b.$$

Similarly, the solution of

$$ya = b \tag{5.3}$$

is unique and is given by

$$y = \frac{1}{< a|a >} b\bar{a}. \tag{5.4}$$

Therefore, we have shown that both real quaternion and octonion algebras are division algebras. We note that complex quaternion and octonion algebras are, however, *not* division algebras, since $< a|a > = \sum_\mu (a_\mu)^2$ could be zero for complex numbers a_μs even though $a \neq 0$. The fact that real para-Hurwitz algebras are also division algebras can be demonstrated in a similar way. Also, if an algebra A is associative with the identity, then a necessary and sufficient condition that A is a division algebra is the existence of the inverse element x^{-1} for any non-zero element $x \in A$.

The next question is whether there are real division algebras whose dimensions are not equal to the canonical values of 1, 2, 4 or 8.

Theorem 5. (Milnor–Bott[32] and Kervair[33]). *The possible dimensions of any real division algebras are limited to only 1, 2, 4, or 8.*

Theorem 6. (Frobenius[34] and Kurosh[35]).

(i) *The possible dimensions of real associative division algebras are limited to 1, 2 and 4, corresponding to real number field, complex number field (regarded as a real algebra of dimension two), and real quaternion algebra (Frobenius).*

(ii) *Any real alternative division algebras are either associative or eight-dimensional octonion algebras (Kurosh).*

Remark 5.1. Theorem 5 by Bott–Milnor and by Kervair has been derived on the basis of topological reasoning on a seven-dimensional sphere. A pure algebraic proof of the theorem is still unknown.

Remark 5.2. A complex algebra is not a division algebra unless it is one-dimensional. For example, complex octonion and para-octonion algebras are not division algebras, since $< x|x >$ could be zero even for non-zero x for the complex cases.

Remark 5.3. The classification of real flexible division algebras has been completed by Benkart and Osborn,[36,37] which, of course, include all real Hurwitz, para-Hurwitz, and pseudo-octonion algebras as special cases.

Remark 5.4. It is also known[7] that any power-associative division algebra must possess the unit element. This fact is an analogue of theorem 4 of Chapter 4.

5.2 Application to real Clifford algebra

We will give an application of the Frobenius theorem on associative division algebras. We note that Dyson[38] has already utilized the theorem for his work on random matrix theory. Following reference 25, we will briefly sketch its application to real matrix representations of Clifford algebras which are associative.

The real Clifford algebra $C(p,q)$ can be defined by the Dirac relation

$$\gamma_\mu \gamma_\nu + \gamma_\nu \gamma_\mu = 2\eta_{\mu\nu} E \qquad (5.5)$$

for $\mu, \nu = 1, 2, \ldots, N$, with

$$N = p + q, \qquad (5.6)$$

where E is the identity matrix and

$$\eta_{\mu\nu} = \begin{cases} 0 & \text{if} \quad \mu \neq \nu \\ 1 & \text{if} \quad \mu = \nu = 1, 2, \ldots, p \\ -1 & \text{if} \quad \mu = \nu = p + 1, \ldots, N. \end{cases} \qquad (5.7)$$

We are concerned here with the real matrix representation of the Dirac matrices γ_μs. Note that the theory of complex matrix realizations is well-known[9] in both physics and mathematics.

Now, 2^N-elements of Clifford algebras are constructed by setting

$$E\ ,\ \gamma_\mu\ ,\ \gamma_\mu \gamma_\nu (\mu < \nu)\ ,\ \gamma_\mu \gamma_\nu \gamma_\lambda (\mu < \nu < \lambda)\ ,\ \ldots\ ,\ \gamma_1 \gamma_2 \ldots \gamma_N\ , \qquad (5.8)$$

which we write collectively as $\Gamma_A (A = 0, 1, 2, \ldots, 2^N - 1)$. Let the γ_μs, and hence the Γ_As, be $d \times d$ irreducible matrices. For any arbitrary $d \times d$ matrix Y, we set

$$S = \sum_A (\Gamma_A)^{-1} Y \Gamma_A . \qquad (5.9)$$

Note that by Eqs. (5.5) and (5.7), we have $(\Gamma_A)^2 = E$ or $-E$, and the inverse $(\Gamma_A)^{-1}$ exists. When we use $\Gamma_A \Gamma_B = \epsilon_{AB} \Gamma_C$ for some Γ_C with $\epsilon_{AB} = \pm 1$, and change the summation variable from Γ_A to Γ_C, we obtain[9] $\Gamma_B^{-1} S \Gamma_B = S$ or

$$S\Gamma_B = \Gamma_B S , \qquad (5.10)$$

for any arbitrary Γ_B. Next, we set

$$M = \{T | T = d \times d \text{ real matrix} ,\ T\Gamma_B = \Gamma_B T \text{ for any } \Gamma_B\} . \qquad (5.11)$$

Clearly, M is a real vector space. Moreover, it becomes a real associative algebra with the unit element E if we define the product in M by the usual matrix product. Next, since the Γ_Bs are assumed to be irreducible, the standard reasoning behind the first Schur lemma implies that any $T \in M$ is an invertible matrix, unless it is identically zero. This will be shown in Remark 5.7 at the end of this section. In other words, M is a real associative division algebra with unit element. Therefore, by the Frobenius theorem, M is isomorphic to either real field, or complex field, or quaternion algebra. This implies that S must be expressed either in the form

(i) $S = a\,E$, for a real constant a

or

(ii) $S = a\,E + b\,J$, for some real constants a and b, where J is a real $d \times d$ matrix satisfying

$$J^2 = -E\,, \tag{5.12a}$$

$$[J, \Gamma_A] = 0\,, \tag{5.12b}$$

or

(iii) $S = a_0 E + a_1 E_1 + a_2 E_2 + a_3 E_3$ for some real constants a_0, a_1, a_2, and a_3, where E_1, E_2, and E_3 are real $d \times d$ matrices satisfying

$$E_j E_k = -\delta_{jk} E + \sum_{\ell=1}^{3} \epsilon_{jk\ell} E_\ell\,, \quad (j, k = 1, 2, 3)\,, \tag{5.13a}$$

$$[E_j, \Gamma_A] = 0\,. \tag{5.13b}$$

Here $\epsilon_{jk\ell}$ is the totally anti-symmetric Levi–Civita symbol in three-dimensional space, with normalization $\epsilon_{123} = 1$. We call these three cases normal, almost complex, and quaternionic respectively. Note that J corresponds to the purely imaginary number $i = \sqrt{-1}$, although J is a real $d \times d$ matrix. Also, for $S = aE + bJ$, the existence of its inverse S^{-1} can be readily verified to be $S^{-1} = (aE - bJ)/(a^2 + b^2)$, just as in the complex numbers. This is the reason it is called an 'almost complex' realization.

At any rate, after some calculations, we can show that the standard orthogonal relation is now replaced by the following.

(i) Normal realization

$$\sum_A \left(\Gamma_A^{-1}\right)_{jk} (\Gamma_A)_{\ell m} = \frac{2^N}{d}\, \delta_{jm}\delta_{\ell k}\,. \tag{5.14}$$

(ii) Almost complex realization

$$\sum_A \left(\Gamma_A^{-1}\right)_{jk} (\Gamma_A)_{\ell m} = \frac{2^N}{d}\, \{\delta_{jm}\delta_{\ell k} - J_{jm}J_{\ell k}\}\,. \tag{5.15}$$

(iii) Quaternionic realization

$$\sum_A \left(\Gamma_A^{-1}\right)_{jk} (\Gamma_A)_{\ell m} = \frac{2^N}{d} \left\{ \delta_{jm}\delta_{\ell k} - \sum_{p=1}^{3}(E_p)_{jm}(E_p)_{\ell k} \right\}. \tag{5.16}$$

In order to derive Eq. (5.15), for example, we used

$$\text{Tr } J = 0, \tag{5.17}$$

which is valid for the following reasons. First, Tr J is a real number, since J is a real matrix. On the other hand, extending the real field into the complex one, we may regard J to be a complex matrix which happens to be real. Then, since the eigenvalues of J must be either i or $-$i by $J^2 = -E$, we see that Tr J must be either 0 or purely imaginary. Combining these statements, we obtain Eq. (5.17).

Next, consider the relation

$$S = aE + bJ = \sum_A (\Gamma_A)^{-1} Y \Gamma_A. \tag{5.18}$$

Taking the trace of both sides, and using Eq. (5.17), we determine a to be

$$a = \frac{1}{d} 2^N \text{ Tr } Y. \tag{5.19a}$$

Similarly, multiplying J by Eq. (5.18) and using

$$[J, \Gamma_A] = 0,$$

we can similarly obtain

$$b = -\frac{1}{d} 2^N \text{ Tr } (JY). \tag{5.19b}$$

Since Y is an arbitrary $d \times d$ real matrix, Eqs. (5.18) and (5.19) now lead to the validity of Eq. (5.15). We can also prove Eq. (5.16) in a similar way.

On the basis of the orthogonality relations (5.14), (5.15), and (5.16), we could derive the following theorem[25] by setting

$$N = p + q = 2n \text{ or } 2n + 1 \tag{5.20}$$

for real irreducible representations of real Clifford algebras: $C(p, q)$.

Theorem 7.

(i) *The normal representation is possible if, and only if, we have*

$$p - q = 0, 1, \text{ or } 2 \pmod 8 \tag{5.21a}$$

with the dimension $d = 2^n$.

(ii) *The almost complex representation is realizable, but only when*

$$p - q = 3 \text{ or } 7 \pmod{8} \tag{5.21b}$$

with $d = 2^{n+1}$.

(iii) *The quaternionic representation is possible if, and only if, we have*

$$p - q = 4, 5, \text{ or } 6 \pmod{8} \tag{5.21c}$$

with $d = 2^{n+1}$.
Moreover, the irreducible representations for the case $N = 2n$ (= even), as well as the almost complex realizations for the case $N = 2n+1 = odd$, are unique. Both normal and quaternionic representations for $N = 2n+1 = odd$ allow two inequivalent irreducible representations with the same dimension, $d = 2^n$ and 2^{n+1}, respectively, which are related to each other by $\tilde{\gamma}_\mu = -\gamma_\mu$.

Proof We will demonstrate only the dimensionality part of the theorem for the case where N = even, leaving the rest to reference 25. For $N = p+q = 2n$ even, it is easy to see that, given $\Gamma_A (\neq E)$, we can always find another Γ_B satisfying

$$\Gamma_A = -\Gamma_B^{-1} \Gamma_A \Gamma_B .$$

For example, for $\Gamma_A = \gamma_1$, we can choose $\Gamma_B = \gamma_2$. This gives

$$\text{Tr } \Gamma_A = -\text{Tr } (\Gamma_B^{-1} \Gamma_A \Gamma_B) = -\text{Tr } \Gamma_A = 0,$$

by the cyclic invariance of the trace, so that

$$\text{Tr } \Gamma_A = \begin{cases} d, & \text{if } \Gamma_A = E \\ 0, & \text{if } \Gamma_A \neq E \end{cases}.$$

Now consider the case of the normal realization. Setting $\ell = m$ and summing over ℓ in Eq. (5.14), gives

$$E_{jk} \text{ Tr } E = \frac{2^N}{d} \delta_{jk} \quad \text{or} \quad d^2 = 2^N = 2^{2n},$$

which gives $d = 2^n$. Similarly, Eq. (5.16) corresponding to the quaternionic realization leads to $d = 2^{n+1}$. However, the case of the almost complex represention, Eq. (5.15), is not possible for $N = 2n$. Setting $\ell = m$, and repeating the same procedure, we now obtain

$$d^2 = 2^{N+1} = 2^{2n+1},$$

which allows no integer solution for d. ∎

Next, we will give as an application the real realization of the quaternion algebra

$$e_j e_k = -\delta_{jk} e_0 + \sum_{\ell=1}^{3} \epsilon_{jk\ell} e_\ell \quad (j,k = 1,2,3),$$

which leads to

$$e_j e_k + e_k e_j = -2\delta_{jk} e_0.$$

Thus, it defines the Clifford algebra $C(0,3)$, which admits only four-dimensional quaternionic representation. Changing notations, we obtain the following real matrix irreducible representation of $C(0,3)$:

$$\gamma_1 = (i\sigma_2) \otimes \sigma_1, \quad \gamma_2 = (i\sigma_2) \otimes \sigma_3, \quad \gamma_3 = E \otimes (i\sigma_2),$$
$$E_1 = \sigma_1 \otimes (i\sigma_2), \quad E_2 = \sigma_3 \otimes (i\sigma_2), \quad E_3 = (i\sigma_2) \otimes E,$$

where E is the 2×2 unit matrix, and σ_1, σ_2, and σ_3 are the familiar 2×2 Pauli matrices

$$\sigma_1 = \begin{pmatrix} 0 & 1 \\ 1 & 0 \end{pmatrix}, \quad \sigma_2 = \begin{pmatrix} 0 & -i \\ i & 0 \end{pmatrix}, \quad \sigma_3 = \begin{pmatrix} 1 & 0 \\ 0 & -1 \end{pmatrix}.$$

Note that two sets of γ_1, γ_2, and γ_3 and of E_1, E_2, and E_3 furnish a four-dimensional real realization of the quaternion algebra, as well as of the Clifford algebra $C(0,3)$. In this connection, we remark that the quaternion algebra does *not* allow any 2×2 real matrix representation, although its 2×2 complex representation is possible, with the usual identification of $E_j = -i\sigma_j$ for $j = 1,2,3$, as we have already remarked in Chapter 1.

Let us return to the discussion of section 3.3. Consider the real octonion algebra. Then, the discussion following Eq. (3.32) now implies that we can normalize the non-zero element b satisfying $< b|b > \neq 0$ only up to signs, that is, $< e_1|e_1 > = \pm 1$. Then we can obtain a basis $e_0(= 1)$, e_1, e_2, ..., e_7 satisfying

$$< e_\mu|e_\nu > = \eta_{\mu\nu} \quad (\mu,\nu = 0,1,2,\dots,7),$$

such that $\eta_{\mu\nu} = 1$ $(\mu = 0,1,2,\dots,p-1)$ and $\eta_{\mu\nu} = -1$ $(\mu = p, p+1,\dots,7)$, where $\eta_{\mu\nu} = 0$ if $\mu \neq \nu$. Therefore, Eq. (3.33) is now replaced by

$$L_j L_k + L_k L_j = -2\eta_{jk} I \quad (j,k = 1,2,\dots,7),$$

which describes the real Clifford algebra $C(q, p-1)$ $(p \geq 1)$ with $q + (p-1) = 7 (= N-1)$. But the L_js are real 8×8 matrices, and hence Theorem 7 requires that this be possible only for the normal representation with $q - (p - 1) = 1$ (mod 8), which allows two solutions $p = 8$ and $q = 0$ or $p = q = 4$. The second solution gives a real octonion algebra which is *not* a division algebra since $< x|x >$ may be zero for non-zero x. The latter algebra can be obtained from the former with the formal replacement of $e_j \rightarrow \sqrt{-1}\, e_j$ $(j = 4, 5, 6, 7)$.

Remark 5.5. From the result of Theorem 7, we can also show that the so-called Majorana–Weyl spinor[39] is possible if, and only if,

$$p = q \text{ (mod 8)}.$$

Note that the Majorana–Weyl spinor is quite important for the so-called super-string theory.[40]

Remark 5.6. The Clifford algebras constructed in Chapters 3 and 4 from real octonions, and real pseudo-octonion algebras, correspond to the real normal irreducible matrix representations of real Clifford algebras $C(0, 7)$ and $C(8, 0)$, respectively, with dimensions 8 and 16.

Remark 5.7. Here, we will demonstrate that the real algebra M defined by Eq. (5.11) is a division algebra. Since M is associative with the unit element, it is sufficient to show that any non-zero T has its inverse T^{-1}. Let V be the underlying real d-dimensional vector space, on which any $T \in M$ operates. Suppose that $T \in M$ is not invertible, that is, T^{-1} does not exist. This implies det $T = 0$ and hence the existence of a real d-dimensional vector $\xi \in V$ $(\xi \neq 0)$ such that

$$T\xi = 0.$$

By multiplying Γ_B and using Eq. (5.11), that is, $T\, \Gamma_B = \Gamma_B T$, we are led to

$$T\, \Gamma_B\, \xi = 0 \tag{5.22}$$

for all Γ_Bs. Let V_0 be a sub-vector space of V consisting of all real linear combinations of $\Gamma_B\, \xi$s, that is, $V_0 = \{x | x = \sum_B C_B \Gamma_B \xi\, , \; C_B \epsilon F\}$, so that $Tx = 0$ for all $x \in V_0$. Note that $V_0 \neq 0$ since $\xi \neq 0$. Moreover, it is an invariant sub-space of V_0, since $\Gamma_A V_0 \subset V_0$ for any Γ_As. However, in view of the irreducibility of the Clifford algebra, this requires that $V_0 = V$, so that $TV_0 = 0$ leads to $T = 0$ identically.

Remark 5.8. We will now explain why similar reasoning does not apply in the case of complex algebra. Suppose that we are now considering the complex representation for the Γ_As. Then M constructed by Eq. (5.11) is now a complex division algebra. However, its dimension is always one, by the following reasoning. The underlying vector space V of Remark 5.7 will now become a complex vector space. Consider the secular equation

$$\det(T - \lambda E) = 0, \tag{5.23}$$

which has at least one solution, λ, which is, in general, a complex number. Then there exists a complex non-zero vector $\xi \in V$ satisfying

$$T \, \xi = \lambda \, \xi.$$

Multiplying Γ_B and using $\Gamma_B T = T \Gamma_B$, we obtain $T(\Gamma_B \xi) = \lambda(\Gamma_B \xi)$, so that

$$Tx = \lambda x \quad \text{for any} \quad x \in V_0. \tag{5.24}$$

The irreducibility for the Clifford algebra again requires that $V_0 = V$, so that Eq. (5.24) now implies the validity of

$$T = \lambda E, \tag{5.25}$$

which gives the usual Schur first lemma for complex irreducible representation. Equation (5.25) also implies Dim $M = 1$. The complex Clifford algebra will give only the normal representation with dimension $d = 2^n$ for $N = 2n$ and $2n + 1$ as in Chapter 3.

This argument can be generalized as follows. Let

$$P(\lambda) = \sum_{n=0}^{N} a_n \lambda^n$$

be a polynomial of the indeterminate λ with $a_n \in F$ for some field F. If the polynomial equation $P(\lambda) = 0$ always has at least one solution for $\lambda \in F$, the field F is termed algebraically closed. The complex field is algebraically closed, while the real field is not. Identifying $P(\lambda) = \det(T - \lambda E)$, the argument presented here is equally applicable to any irreducible representation in any algebraically closed field F.

Remark 5.9. More recently, Goddard *et al.*[41] have given another application of real unital division algebra in connection with the construction of the vertex operators in the string model, where they note that the operator product expansion of the fermions is related in a precise way to one or other of the division algebras given by the complex numbers, quaternions, or octonions.

5.3 Dimensional regularization and non-associative Dirac algebra

As is well-known, one complication when dealing with a quantum field theory such as QCD (Quantum Chromo Dynamics) is the necessity of renormalization. A simple way of dealing with this problem is the dimensional regularization method,[42] in which we calculate Feynmann integrals in an arbitrary real dimension N, and take the limit $N \to 4$ after the end of the calculation. Then the divergences will appear as poles at $N = 4$, and can be readily disposed of by renormalization. Although the method works well in general, a complication will occur in the calculation of the so-called triangle anomaly, where we compute a triangle fermion loop. Since the absence of the triangle anomaly is crucial for the renormalizeability of any gauge field theory, the problem has been studied by many authors.[43]

The source of the complication arises in connection with the problem of how we define the so-called Γ_5 matrix. The standard Clifford algebra in N dimensions is associative and is generated by N elements $\gamma_\mu (\mu = 1, 2, \dots, N)$ by Eq. (5.5). For the physical space-time dimension of $N = 4$, we normally define the element Γ_5 by

$$\Gamma_5 = \sum_{\mu, \nu, \alpha, \beta = 1}^{4} \frac{1}{4!} \, \epsilon^{\mu\nu\alpha\beta} \gamma_\mu \gamma_\nu \gamma_\alpha \gamma_\beta = \gamma_1 \gamma_2 \gamma_3 \gamma_4 , \qquad (5.26)$$

in terms of the Levi–Civita symbol $\epsilon^{\mu\nu\alpha\beta}$ in four dimensions. For the dimensional regularization method, we must define the Γ_5 operation, even in the N-dimensional case, by assuming that E, γ_μ, $\gamma_\mu\gamma_\nu (\mu < \nu)$, $\gamma_\mu\Gamma_5$, and Γ_5 for $\mu, \nu = 1, 2, \dots, N$ obey a multiplication table which mimics that of the $N = 4$ Clifford–Dirac algebra. Note that the notation $\epsilon_{\mu\nu\alpha\beta} (\mu, \nu, \alpha, \beta = 1, 2, \dots, N)$ is now a totally anti-symmetric numerical tensor which reduces to the Levi–Civita symbol for the limit $N \to 4$.

Following reference 44, the desired multiplication table for the algebra A is now given by

(i) $1x = x1 = x \quad$ for $\quad x \in A,$ $\qquad\qquad$ (5.27a)

(ii) $\gamma_\mu\gamma_\nu + \gamma_\nu\gamma_\mu = 2 \, \eta_{\mu\nu} \, 1,$ $\qquad\qquad$ (5.27b)

(iii) $\gamma_\mu\Gamma_5 + \Gamma_5\gamma_\mu = 0,$ $\qquad\qquad$ (5.27c)

(iv) $\Gamma_5\Gamma_5 = -1,$ $\qquad\qquad$ (5.27d)

(v) $\Gamma_5(\Gamma_5\gamma_\mu) = (\gamma_\mu\Gamma_5)\Gamma_5 = -\gamma_\mu,$ $\qquad\qquad$ (5.27e)

(vi) $(\Gamma_5\gamma_\mu)(\Gamma_5\gamma_\nu) = \gamma_\mu\gamma_\nu,$ $\qquad\qquad$ (5.27f)

(vii) $\gamma_\lambda(\gamma_\mu\gamma_\nu) = (\gamma_\lambda\gamma_\mu)\gamma_\nu = \eta_{\mu\nu}\gamma_\lambda + \eta_{\lambda\mu}\gamma_\nu - \eta_{\lambda\nu}\gamma_\mu - \epsilon_{\lambda\mu\nu\alpha}\gamma^\alpha\Gamma_5,$ $\quad$ (5.27g)

(viii) $\Gamma_5(\gamma_\mu \gamma_\nu) = (\gamma_\mu \gamma_\nu)\Gamma_5 = \epsilon_{\mu\nu\alpha\beta}\gamma^\alpha\gamma^\beta$,   (5.27h)

(ix) $(\Gamma_5\gamma_\mu)\gamma_\nu = -\gamma_\mu(\Gamma_5\gamma_\nu) = \eta_{\mu\nu}\Gamma_5 - \frac{1}{2}\,\epsilon_{\mu\nu\alpha\beta}\gamma^\alpha\gamma^\beta$,   (5.27i)

(x) $(\Gamma_5\gamma_\lambda)(\gamma_\mu\gamma_\nu) = \eta_{\mu\nu}\Gamma_5\gamma_\lambda + \eta_{\lambda\mu}\Gamma_5\gamma_\nu - \eta_{\lambda\nu}\Gamma_5\gamma_\mu - 2\epsilon_{\lambda\mu\nu\alpha}\gamma^\alpha$,   (5.27j)

(xi) $(\gamma_\mu\gamma_\nu)(\Gamma_5\gamma_\lambda) = \eta_{\mu\nu}\Gamma_5\gamma_\lambda - \eta_{\lambda\mu}\Gamma_5\gamma_\nu + \eta_{\lambda\nu}\Gamma_5\gamma_\mu - 2\epsilon_{\lambda\mu\nu\alpha}\gamma^\alpha$,   (5.27k)

(xii) $[\gamma_\alpha, \gamma_\beta][\gamma_\mu, \gamma_\nu] = 2\{\eta_{\beta\mu}[\gamma_\alpha, \gamma_\nu] + \eta_{\alpha\nu}[\gamma_\beta, \gamma_\mu] - \eta_{\beta\nu}[\gamma_\alpha, \gamma_\mu] - \eta_{\alpha\mu}[\gamma_\beta, \gamma_\nu]\}$
$$+ 4(\eta_{\beta\mu}\eta_{\alpha\nu} - \eta_{\beta\nu}\eta_{\alpha\mu})1 + 4\epsilon_{\alpha\beta\mu\nu}\Gamma_5 . \qquad (5.27l)$$

Here we have replaced E by 1, and the repeated Greek indices imply automatic summation over N-values $1, 2, \ldots, N$. Further, $[\gamma_\mu, \gamma_\nu]$ in Eq. (5.27ℓ) is the commutator

$$[\gamma_\mu, \gamma_\nu] = \gamma_\mu\gamma_\nu - \gamma_\nu\gamma_\mu . \qquad (5.28)$$

Note that the dimension of the present algebra is

$$\text{Dim } A = \frac{1}{2}\,(N^2 + 3N + 4) \qquad (5.29)$$

instead of 2^N.

Except for the case where $N = 4$, our algebra A is no longer associative. The reason is that the familiar formula

$$\epsilon_{\lambda\mu\nu\alpha}\eta^{\alpha\beta}\epsilon_{\tau\rho\theta\beta} = \eta_{\lambda\tau}(\eta_{\mu\rho}\eta_{\nu\theta} - \eta_{\nu\rho}\eta_{\mu\theta})$$
$$- \eta_{\lambda\rho}(\eta_{\mu\tau}\eta_{\nu\theta} - \eta_{\nu\tau}\eta_{\mu\theta}) - \eta_{\lambda\theta}(\eta_{\mu\rho}\eta_{\nu\tau} - \eta_{\nu\rho}\eta_{\mu\tau})$$

is no longer valid except in the special case where $N = 4$. Indeed, we find, for example, a non-zero expression for the associator $(\gamma_\mu\Gamma_5, \gamma_\nu, \Gamma_5)$, as is computed in reference 44. The algebra A is also neither flexible, nor power-associative, but it is Lie-admissible. If we set

$$J_{\mu\nu} = -J_{\nu\mu} = \frac{1}{4}\,[\gamma_\mu, \gamma_\nu], \qquad (5.30a)$$

$$J_{\mu, N+1} = -J_{N+1, \mu} = \frac{1}{2}\,\gamma_\mu, \qquad (5.30b)$$

$$J_{\mu, N+2} = -J_{N+2, \mu} = \frac{1}{2}\,\gamma_\mu\Gamma_5, \qquad (5.30c)$$

$$J_{N+1, N+2} = -J_{N+2, N+1} = \frac{1}{2}\,\Gamma_5, \qquad (5.30d)$$

as well as

$$\eta_{N+1, N+1} = \eta_{N+2, N+2} = -1, \quad \eta_{N+1, N+2} = 0, \qquad (5.31a)$$
$$\eta_{\mu, N+1} = \eta_{\mu, N+2} = 0, \qquad (5.31b)$$

for $\mu, v = 1, 2, \ldots, N$, then we obtain the $so(N+2)$ Lie algebra relation

$$[J_{ab}, J_{cd}] = \eta_{bc}J_{ad} - \eta_{ac}J_{bd} - \eta_{bd}J_{ac} + \eta_{ad}J_{bc} \qquad (5.32)$$

for all $a, b, c, d = 1, 2, \ldots, N+2$.

Moreover, we can define an invariant symmetric trace form. We set

$$\mathrm{Tr}\ 1 = N, \quad \mathrm{Tr}\ (\gamma_\mu \gamma_v) = \eta_{\mu v} N, \qquad (5.33a)$$
$$\mathrm{Tr}\ x = 0 \quad \text{for} \quad x = \Gamma_5, \Gamma_5\gamma_\mu, \gamma_\mu. \qquad (5.33b)$$

Then we can verify the validity of

$$\mathrm{Tr}\ xy = \mathrm{Tr}\ yx, \qquad (5.54a)$$
$$\mathrm{Tr}\ \{(xy)z\} = \mathrm{Tr}\ \{x(yz)\}, \qquad (5.34b)$$

so that if we set

$$< x|y > = \mathrm{Tr}\ xy, \qquad (5.35)$$

it defines a non-degenerate bi-linear symmetric form satisfying

$$< xy|z > = < x|yz > . \qquad (5.36)$$

Moreover, the algebra satisfies other interesting properties, which can be found in reference 44.

Remark 5.10. The definition of Γ_5 here differs from that in reference 44 by the factor i.

6

Clebsch–Gordan algebras

6.1 SU(2) case

For many physicists, non-associative algebras may appear exotic and remote from real physics. However they already occur in the context of angular momentum, familiar in quantum mechanics. Let $\mathbf{J}_1$, $\mathbf{J}_2$, and $\mathbf{J}_3$ be the angular momentum operators of three particles. (In reality, we consider their representations.) It is well known that their sum depends upon how we proceed, that is, we have $(\mathbf{J}_1 + \mathbf{J}_2) + \mathbf{J}_3 \neq \mathbf{J}_1 + (\mathbf{J}_2 + \mathbf{J}_3)$, in some sense. We can depict the situation graphically as in Fig. 6.1.

But, from the mathematical viewpoint, the resulting non-associative algebra is *not* so interesting for the most general case, since it is very complicated and infinite dimensional. However, if we project some particular angular momentum states only, then we may obtain some interesting results. First let $\{j\}$ for $j = 0$, $\frac{1}{2}$, 1, $\frac{3}{2}$, 2, ... be the $2j + 1$-dimensional irreducible space of the group SU(2), and note the familiar Clebsch–Gordan decomposition rule[45]

$$\{j_1\} \otimes \{j_2\} = \{j_1 + j_2\} \oplus \{j_1 + j_2 - 1\} \oplus \ldots \oplus \{|j_1 - j_2|\}. \qquad (6.1)$$

Let $\begin{pmatrix} j_1 & j_2 & j_3 \\ m_1 & m_2 & m_3 \end{pmatrix}$ be Wigner's 3-j symbol[45,46] for the recoupling of two angular momenta j_1 and j_2 into j_3. We consider here only the case where $J = j_1 = j_2 = j_3$ is a positive integer. Then the state vectors

$$\psi_m = |J, m> = Y_{Jm}(\theta, \phi) \qquad (6.2)$$

for $m = J$, $J - 1$, ..., $-(J - 1)$, $-J$ form a basis of the $2J + 1$ vector space $V = \{J\}$ with the orthogonality relation

$$< \psi_m | \psi_{m'} > = \frac{1}{2} (-1)^{m+1} \delta_{m+m',0}. \qquad (6.3)$$

64

Fig. 6.1 Non-associativity for angular momenta.

We should warn readers that $< x|y > = < y|x >$ here is bi-linear in x and y but *not* sesquilinear, that is, $< x|y > \neq (< y|x >)^*$ if F is a complex field. In particular, $< x|y >$ $(x, y \in V)$ defines a bi-linear symmetric non-degenerate form in V. Also, the unconventional factor $\frac{1}{2}$ in Eq. (6.3) is for later convenience. We can now introduce the su(2)-invariant product $\psi_{m_1} \cdot \psi_{m_2}$ in V by

$$\psi_{m_1} \cdot \psi_{m_2} = a_J \sum_{m=-J}^{J} \begin{pmatrix} J & J & J \\ m_1 , & m_2 , & m \end{pmatrix} (-1)^m \psi_{-m}, \qquad (6.4)$$

where a_J is a suitable normalization constant. We now note[45,46] that

$$\begin{pmatrix} J & J & J \\ m_1 & m_2 & m_3 \end{pmatrix}$$

is either totally symmetric or totally anti-symmetric, respectively, in m_1, m_2, and m_3, depending upon whether $J =$ even or $J =$ odd. As we will see shortly, two cases of $J = 1$ and $J = 3$ are of some particular interest, where Eq. (6.4) defines a Lie algebra SO(3) for $J = 1$ and the so-called seven-dimensional exceptional Malcev algebra[22] for $J = 3$. However, a slightly more interesting case can be obtained by adding a spin 0 (i.e. $J = 0$) singlet term e_0 into the system, and by modifying Eq. (6.4) into

$$\psi_{m_1} \cdot \psi_{m_2} = -b_J < \psi_{m_1}|\psi_{m_2} > e_0 + a_J \sum_{m=-J}^{J} \begin{pmatrix} J & J & J \\ m_1 , & m_2 , & m \end{pmatrix} (-1)^m \psi_{-m},$$
$$(6.5)$$

with the following additional assumption:

$$\psi_{m_1} \cdot e_0 = e_0 \cdot \psi_{m_1} = \pm \psi_{m_1}, \qquad (6.6a)$$

$$e_0 \cdot e_0 = e_0. \qquad (6.6b)$$

By choosing arbitrary constants b_J and a_J suitably, Eqs. (6.5) and (6.6) now realize the quaternion (or para-quaternion) for $J = 1$, and octonion (or para-octonion) algebras for $J = 3$, depending upon the two possible signs in Eq. (6.6a).

Let us demonstrate these statements. For $J = 1$, it is easier to deal with the Cartesian vectors e_1, e_2, and e_3 given by

$$\psi_0 = e_3 , \quad \psi_1 = \frac{-1}{\sqrt{2}} (e_1 - ie_2) , \quad \psi_{-1} = \frac{1}{\sqrt{2}} (e_1 + ie_2) .$$

When we choose constants a_1 and b_1 suitably, then Eq. (6.5) can easily be shown to become

$$e_j \cdot e_k = -\delta_{jk} e_0 + \sum_{\ell=1}^{3} \epsilon_{jk\ell} e_\ell ,$$

while Eqs. (6.6) become $e_j \cdot e_0 = e_0 \cdot e_j = \pm e_j$ and $e_0 \cdot e_0 = e_0$. (Here and hereafter in this section, the symbol $x \cdot y$ has nothing to do with the Jordan product defined by Eq. (2.16b).) These are precisely the multiplication tables of quaternion or para-quaternion algebras, depending upon the two signs. Alternatively, if we choose $b_1 = 0$, it gives an algebra isomorphic to the original su(2) Lie algebra.

For the case where $J = 3$, we choose $a_3 = \sqrt{21}$ and set

$$u_1 = \psi_1 , \quad u_2 = \psi_2 , \quad u_3 = \psi_{-3} ,$$
$$\bar{u}_1 = \psi_{-1} , \quad \bar{u}_2 = -\psi_{-2} , \quad \bar{u}_3 = \psi_3 , \tag{6.7}$$

then Eq. (6.4), for example, reproduces the following multiplication table:

$$u_j \cdot u_k = \sum_{\ell=1}^{3} \epsilon_{jk\ell} \bar{u}_\ell ,$$
$$\bar{u}_j \cdot \bar{u}_k = \sum_{\ell=1}^{3} \epsilon_{jk\ell} u_\ell ,$$
$$u_j \cdot \bar{u}_k = -\bar{u}_k \cdot u_j = -\frac{1}{\sqrt{2}} \delta_{jk} \psi_0 , \tag{6.8}$$
$$\psi_0 \cdot u_j = -u_j \cdot \psi_0 = \frac{1}{\sqrt{2}} u_j ,$$
$$\psi_0 \cdot \bar{u}_j = -\bar{u}_j \cdot \psi_0 = -\frac{1}{\sqrt{2}} \bar{u}_j ,$$

with

$$< \bar{u}_j | u_k > \, = \, < u_k | \bar{u}_j > \, = \frac{1}{2} \delta_{jk} ,$$
$$< \psi_0 | \psi_0 > \, = -\frac{1}{2} , \tag{6.9}$$

while all other $< x|y >$ are identically zero. We note that Eq. (6.8) defines the seven-dimensional exceptional Malcev algebra, as is discussed in the following section.

Next, we consider the more interesting case of Eq. (6.5) by adjoining the unit element e_0. Then, setting $b_3 = 1$ with the choice of the upper sign in Eq. (6.6a), and introducing u_0 and $\overline{u}_0$ by

$$u_0 = \frac{1}{2} \left(e_0 + \sqrt{2} \, \psi_0 \right) , \quad \overline{u}_0 = \frac{1}{2} \left(e_0 - \sqrt{2} \, \psi_0 \right) , \qquad (6.10)$$

we obtain

$$u_j \cdot u_k = \sum_{\ell=1}^{3} \epsilon_{jk\ell} \overline{u}_\ell ,$$

$$\overline{u}_j \cdot \overline{u}_k = \sum_{\ell=1}^{3} \epsilon_{jk\ell} u_\ell ,$$

$$u_j \cdot \overline{u}_k = -\delta_{jk} u_0 ,$$

$$\overline{u}_k \cdot u_j = -\delta_{jk} \overline{u}_0 ,$$

$$u_0 \cdot u_j = u_j \cdot \overline{u}_0 = u_j , \qquad (6.11)$$

$$\overline{u}_0 \cdot \overline{u}_j = \overline{u}_j \cdot u_0 = \overline{u}_j ,$$

$$u_j \cdot u_0 = \overline{u}_j \cdot \overline{u}_0 = \overline{u}_0 \cdot u_j = u_0 \cdot \overline{u}_j = 0 ,$$

$$u_0 \cdot u_0 = u_0 ,$$

$$\overline{u}_0 \cdot \overline{u}_0 = \overline{u}_0 ,$$

$$u_0 \cdot \overline{u}_0 = \overline{u}_0 \cdot u_0 = 0 .$$

If we set

$$u_1 = \frac{1}{2} \left(e_1 + i e_4 \right) , \quad u_2 = \frac{1}{2} \left(e_2 + i e_5 \right) , \quad u_3 = \frac{1}{2} \left(e_3 + i e_6 \right) ,$$

$$\overline{u}_1 = \frac{1}{2} \left(e_1 - i e_4 \right) , \quad \overline{u}_2 = \frac{1}{2} \left(e_2 - i e_5 \right) , \quad \overline{u}_3 = \frac{1}{2} \left(e_3 - i e_6 \right) , \qquad (6.12)$$

$$u_0 = \frac{1}{2} \left(e_0 - i e_7 \right) , \quad \overline{u}_0 = \frac{1}{2} \left(e_0 + i e_7 \right) ,$$

then Eq. (6.11) will reproduce the octonionic multiplication table, Eq. (1.31), with $e_j \cdot e_k \equiv e_j e_k$ in terms of $e_0, e_1, e_2, \ldots, e_7$, where e_0 is the unit element. In this connection, we would like to mention that, in contrast, we would obtain a para-octonion algebra if we were to adopt the lower negative sign in the right side of Eq. (6.6a).

The reason that the case of $J = 3$ recouplings leads to octonion algebras will be explained in Section 6.3 below.

Remark 6.1. Two eight-dimensional algebras spanned by e_μ ($\mu = 0, 1, 2, \ldots, 7$) and u_j, $\bar{u}_j$, u_0 and $\bar{u}_0$ ($j = 1, 2, 3$) give the same complex octonion algebras, since two bases are related to each other by Eq. (6.12). However, regarded as *real* algebras, they lead to two inequivalent real octonion algebras for the following reasons. As we have noted already, the real octonion algebra defined in terms of e_0, e_1, e_2, $\ldots$, e_7 is a real division algebra. However, the real algebra (known as the split Cayley algebra) defined by Eq. (6.11) is *not* a real division algebra. In fact, consider equation $u_0 \cdot x = 0$, whose solution x is not unique. Indeed, choices of $x = 0$ or $x = \bar{u}_0$ or, more generally, $x = \lambda \bar{u}_0$ for arbitrary constant λ are all solutions. This is, of course, due to the fact that $< u_0 | u_0 > = 0$, violating the condition of real division algebra discussed in Chapter 5.

Remark 6.2. The case of $J = 2$ gives the commutative algebra. This algebra, together with the addition of the $J = 0$ component can be shown to lead to a six-dimensional special Jordan algebra, although we will not go into the details here.

6.2 Homomorphism ($W_1 \to W_2$)

As an illustration, consider again the su(2) case with the product defined by Eq. (6.4). We note that Eqs. (6.3) and (6.4) are invariant by construction under any SO(3) transformation. In terms of the Lie algebra SO(3) = su(2), this implies the validity of

$$< b\psi_{m_1} | \psi_{m_2} > \; + < \psi_{m_1} | b\psi_{m_2} > = b < \psi_{m_1} | \psi_{m_2} > = 0 \,, \quad (6.13)$$

$$b\big(\psi_{m_1} \cdot \psi_{m_2}\big) = \big(b\psi_{m_1}\big) \cdot \psi_{m_2} + \psi_{m_1} \cdot \big(b\psi_{m_2}\big) \,, \quad (6.14)$$

under the actions of any $b \in$ su(2) (i.e. $b = J_3$ or $J_\pm$). For simplicity, let V be the $2J+1$ dimensional space $\{J\}$ spanned by ψ_m ($m = J, J-1, \ldots, -J$). With further restriction to cases of J being odd integers, we can interpret the anti-symmetric product $x \cdot y$ as a mapping

$$f \; : \; (V \otimes V)_A \to V \,, \quad (6.15a)$$

with

$$f((x \otimes y)_A) = x \cdot y = -y \cdot x \,, \quad (6.15b)$$

while $< x | y >$ may be identified by another mapping

$$\phi \; : \; (V \otimes V)_S \to F \,, \quad (6.16a)$$

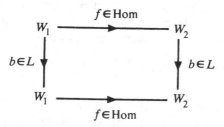

Fig. 6.2 Commutativity law $b \circ f = f \circ b$.

such that

$$\phi((x \otimes y)_S) = <x|y> = <y|x> . \tag{6.16b}$$

Here, the suffixes A and S designate the anti-symmetric and symmetric components of the tensor product $V \otimes V$. Moreover, conditions (6.13) and (6.14) can then be interpreted to imply the commutativities of the action b of the su(2) with mappings ϕ and f, that is,

$$\begin{aligned} b \circ \phi = \phi \circ b \, , \\ b \circ f = f \circ b \, , \end{aligned} \tag{6.17}$$

where we use

$$b(x \otimes y) = (bx) \otimes y + x \otimes (by) \tag{6.18}$$

for the action of $b \, \epsilon$ su(2) on $V \otimes V$. (Note that $b = \mathbf{J}^{(1)} + \mathbf{J}^{(2)}$ in physics notation.) Here $b \circ \phi$, for example, designates the successive composition map of ϕ followed by b. We then write

$$\begin{aligned} \phi \, \epsilon \, \text{Hom} \, ((V \otimes V)_S \rightarrow F) \, , \\ f \, \epsilon \, \text{Hom} \, ((V \otimes V)_A \rightarrow V) \, , \end{aligned}$$

where Hom $(W_1 \rightarrow W_2)$ implies the vector spaces of all homomorphisms between su(2) representation spaces W_1 and W_2, which commute with the actions of any $b \, \epsilon$ su(2).

Now the generalization for any simple (or semi-simple) Lie algebra L is obvious. Let W_1 and W_2 be two L-modules, that is, representation spaces of L, which may not necessarily be irreducible. Then

$$f \, \epsilon \, \text{Hom} \, (W_1 \rightarrow W_2) \tag{6.19}$$

is a linear mapping of W_1 into W_2 which commutes with $b \, \epsilon \, L$, that is,

$$b \circ f = f \circ b \, . \tag{6.20}$$

See Fig. 6.2 for a graphical representation of this equation.

Let W_2 be an irreducible module of a simple (or semi-simple) Lie algebra L. Any L-module W_1 on the other side can be decomposed as a direct sum of inequivalent irreducible spaces V_j as

$$W_1 = \sum_j \oplus m_j V_j \,, \qquad (6.21)$$

where m_j is the multiplicity of the irreducible space V_j. We now note the validity of the following theorem,[47] which is actually a special case of the Wigner–Eckart theorem in physics.[46]

Theorem 8. *Let Dim Hom $(W_1 \rightarrow W_2)$ be the dimension of the vector space Hom $(W_1 \rightarrow W_2)$. Suppose that V_k is the kth irreducible space of W_1 as in Eq. (6.21), which is isomorphic to the irreducible L-module W_2. We then have*

$$\text{Dim Hom } (W_1 \rightarrow V_k) = m_k \ .$$

Here, we regard the underlying field F to be the one-dimensional trivial representation $\{0\}$ of L.

An application of Theorem 8 for the special case of $L = su(2)$ is

$$\text{Dim Hom } (\{j_1\} \otimes \{j_2\} \rightarrow \{j\}) = \begin{cases} 1 \,, & \text{if } j_1 + j_2 \geq j \geq |j_1 - j_2| \\ 0 \,, & \text{otherwise} \end{cases}$$

by the Clebsch–Gordan decomposition, Eq. (6.1).

Returning to the general case, we fix the irreducible L-module V and decompose the tensor product

$$V^n = V \otimes V \otimes V \otimes \ldots \otimes V \qquad (n \text{ times})$$

into

$$V^n = \sum_{(f_1, f_2 \ldots f_n)} \oplus \, [f_1, f_2, \ldots, f_n] \,,$$

where $[f_1, f_2, \ldots, f_n]$ designates the Young-tableau symbol[48,49] satisfying

$$f_1 \geq f_2 \geq \ldots \geq f_n \geq 0$$

and

$$f_1 + f_2 + \ldots + f_n = n \ .$$

As usual, we set

$$V = [1] \quad \text{and} \quad F = [0] \qquad (6.22)$$

and for the example $n = 3$ we write

$$[1^3] = [1, 1, 1] \ ,$$
$$[3] = [3, 0, 0] \ ,$$
$$[2, 1] = [2, 1, 0] \ ,$$

as usual. Then $(V \otimes V)_S = [2, 0] \equiv [2]$ and $(V \otimes V)_A = [1, 1] = [1^2]$ for $n = 2$.

Following the method described in reference 50, we first assume that we have

$$\text{Dim Hom } ([2] \rightarrow F) = 1 \ . \tag{6.23}$$

This implies the existence of the symmetric bi-linear form $< x|y > = < y|x >$, which is unique, apart from the overall normalization. Moreover, it is invariant under the actions of $b \ \epsilon \ L$, that is,

$$< bx|y > \ + \ < x|by > = 0 \ , \tag{6.24}$$

by construction as in Eq. (6.13). Since the space V is irreducible, the standard reasoning analogous to Schur's lemma then requires the non-degeneracy of $< x|y >$. We also assume the validity of

$$\text{Dim Hom } ([1^2] \rightarrow [1]) = 1 \ , \tag{6.25}$$

as well as

$$\text{Dim Hom } ([1^3] \rightarrow F) = 1 \ . \tag{6.26}$$

Consider Eq. (6.25), which implies the existence of the unique (apart from the normalization constant) anti-symmetric product

$$x \cdot y = -y \cdot x \tag{6.27}$$

in V, while Eq. (6.26) requires the existence of a totally anti-symmetric triple-linear function $\psi(x, y, z) \ \epsilon \ F$. By using the non-degeneracy of $< z|x >$, we can always write any triple-linear function as

$$\psi(x, y, z) = \ < z|w(x, y) > \ ,$$

for some $w(x, y) = -w(y, x) \ \epsilon \ V$, which must be bi-linear in x and y. However, the uniqueness of the bi-linear anti-symmetric product $x \cdot y$ in V requires that $w(x, y) = x \cdot y$, apart from a constant. Therefore, we conclude that

$$< z|x \cdot y > \ = \ \text{totally anti-symmetric in } x, y, z \ , \tag{6.28a}$$

or

$$< z|x \cdot y > \; = \; < z \cdot x|y > \; . \tag{6.28b}$$

Finally, in order to obtain the octonionic structure, it is necessary to make the last assumption of

$$\text{Dim Hom } ([2,1] \rightarrow [1]) = 1 \; . \tag{6.29}$$

In order to apply Eq. (6.29), let us consider the following two maps, f_1 and f_2:

$$V \otimes V \otimes V \rightarrow V \; ,$$

defined by

$$f_1(x,y,z) = z \cdot (x \cdot y) + x \cdot (z \cdot y) \tag{6.30a}$$

$$f_2(x,y,z) = 2 < x|z > y \; - \; < y|z > x \; - \; < y|x > z \; . \tag{6.30b}$$

We can readily verify the following conditions:

$$(0) \quad b \circ f_j = f_j \circ b \quad \text{for any} \quad b \in L \; , \tag{6.31a}$$

$$(1) \quad f_j(x,y,z) = f_j(z,y,x) \; , \tag{6.31b}$$

$$(2) \quad f_j(x,y,z) + f_j(y,z,x) + f_j(z,x,y) = 0 \; , \tag{6.31c}$$

for $j = 1, 2$. In other words, these conditions imply

$$f_1, f_2 \in \text{Hom } ([2,1] \rightarrow [1]) \; . \tag{6.32}$$

But by Eq. (6.29), f_1 and f_2 must be linearly dependent. In other words, there exist constants α and β, at least one of which is non-zero, such that

$$\alpha f_1 + \beta f_2 = 0 \; .$$

However, we must have $\alpha \neq 0$ since if $\alpha = 0$, then $\beta \neq 0$ and hence $f_2 = 0$ identically. But the equation $f_2(x,y,z) = 0$ is satisfied only for the trivial case of Dim $V = 1$. Therefore, setting $\lambda_0 = -\frac{\beta}{\alpha}$, we conclude that

$$f_1(x,y,z) = \lambda_0 f_2(x,y,z)$$

or

$$z \cdot (x \cdot y) + x \cdot (z \cdot y) = \lambda_0 \{2 < x|z > y \; - \; < y|z > x \; - \; < y|x > z\} \tag{6.33}$$

for some constants λ_0. Setting $z = x$, Eq. (6.33) is equivalent to

$$x \cdot (x \cdot y) = \lambda_0 \{< x|x > y \; - \; < y|x > x\} \; . \tag{6.34}$$

From Eq. (6.34), we can prove the validity of the Malcev identity[22]

$$(x \cdot y) \cdot (x \cdot z) = \{(x \cdot y) \cdot z\} \cdot x + \{(y \cdot z) \cdot x\} \cdot x + \{(z \cdot x) \cdot x\} \cdot y \; , \tag{6.35}$$

so that the present product $x \cdot y = -y \cdot x$ defines a Malcev algebra.

Returning to the discussion of Eq. (6.34), it is almost equivalent to one of the alternative laws of Eq. (3.18) when we interchange the roles of x and y. Actually, we can convert the present algebra into a Hurwitz algebra if we adjoin[50] the unit element e by extending our vector space V into

$$V^* = V \oplus Fe . \qquad (6.36)$$

Let X and $Y \in V^*$ be two generic elements of V^* given by

$$\begin{aligned} X &= x + \mu e , \quad (x \in V , \ \mu \in F) , \\ Y &= y + v e , \quad (y \in V , \ v \in F) . \end{aligned} \qquad (6.37)$$

We define the product XY and symmetric bi-linear form $< X|Y >$ in V^* by

$$XY = x \cdot y + \mu y + v x + \{\mu v + \lambda_0 < x|y >\}e , \qquad (6.38a)$$

$$< X|Y > = \mu v - \lambda_0 < x|y > . \qquad (6.38b)$$

Moreover, setting

$$\overline{X} = 2 < X|e > e - X = \mu e - x , \qquad (6.39)$$

we can prove that the new product satisfies the following relations:

$$\begin{aligned} &(0) \quad Xe = eX = X , \\ &(i) \quad X\overline{X} = \overline{X}X = < X|X > e , \\ &(ii) \quad \overline{XY} = \overline{Y}\,\overline{X} , \\ &(iii) \quad < XY|Z > = < X|Z\overline{Y} > , \\ &(iv) \quad < XY|XY > = < X|X >< Y|Y > , \\ &(v) \quad X(\overline{X}Y) = (X\overline{X})Y = < X|X > Y , \\ &(vi) \quad (YX)\overline{X} = Y(X\overline{X}) = < X|X > Y . \end{aligned} \qquad (6.40)$$

Finally, we can easily verify the non-degeneracy of $< X|Y >$, provided that we have $\lambda_0 \neq 0$. Assuming this, we see that the present algebra defines the Hurwitz algebra and hence Dim $V^* = $ Dim $V + 1$ is possible only for values of 1, 2, 4, or 8.

Remark 6.3. The Lie algebra L is clearly a part of the derivation algebra of the present algebra by construction. The case of $L = $ su(2) with V having three-dimensional vector space gives the quaternion algebra for V^*.

Remark 6.4. For the Hurwitz algebra V^*, we define its commutator algebra by

$$[X, Y] = XY - YX = x \cdot y - y \cdot x = 2x \cdot y,$$

so that it defines a Malcev-algebra. In such a case we call V^* a Malcev-admissible algebra.[22] Thus, octonion algebra is Malcev-admissible.

6.3 G_2 and split-Cayley algebra

One interesting application of the results given in section 6.2 is obtained when we identify V to be the seven-dimensional irreducible representation of the Lie algebra G_2. Since G_2 is of rank 2, any irreducible representation $\{\Lambda\}$ of G_2 is specified by its highest weight Λ in the form

$$\Lambda = m_1\Lambda_1 + m_2\Lambda_2$$

in terms of two non-negative integers m_1 and m_2, where Λ_1 and Λ_2 are two fundamental weights of G_2. Adopting the lexicographical ordering of simple roots of G_2 as in reference 51, $V = [1]$ may be identified with $V = \{\Lambda_2\}$. Utilizing the method of general Dynkin indices given elsewhere,[52] we can then verify the validity of

$$[2] = \{0\} \oplus \{2\Lambda_2\} \quad \text{or} \quad 28 = 1 + 27,$$
$$[1^2] = [1] \oplus \{\Lambda_1\} \quad \text{or} \quad 21 = 7 + 14,$$
$$[1^3] = \{0\} \oplus [1] \oplus \{2\Lambda_2\} \quad \text{or} \quad 35 = 1 + 7 + 27,$$

as well as

$$[2, 1] = [1] \oplus \{\Lambda_1\} \oplus \{2\Lambda_2\} \oplus \{\Lambda_1 + \Lambda_2\},$$

or

$$112 = 7 + 14 + 27 + 64.$$

Therefore, by Theorem 8, we find the validity of

$$\text{Dim Hom } ([2] \rightarrow F) = 1,$$
$$\text{Dim Hom } ([1^2] \rightarrow [1]) = 1,$$
$$\text{Dim Hom } ([1^3] \rightarrow F) = 1,$$
$$\text{Dim Hom } ([2, 1] \rightarrow [1]) = 1,$$

so that we can construct the octonion algebra from the seven-dimensional irreducible representation of G_2, by the results of the preceding section. However, the explicit construction of its multiplication table depends upon the choice of the basis of the module.

(a) su(3) basis

We first choose a basis of G_2 as follows. We know that G_2 is a 14-dimensional Lie algebra, and that it contains the eight-dimensional su(3) as its maximal sub-Lie algebra. If we restrict G_2 to su(3), then the 14-dimensional adjoint representation of G_2 will reduce to[51]

$$14 \rightarrow 14 = 8 \oplus 3 \oplus \bar{3} \ .$$

Therefore, we can label a basis of G_2 by su(3) tensors B_ν^μ, A_ν, and A^μ ($\mu, \nu = 1, 2, 3$) satisfying

$$\sum_{\mu=1}^{3} B_\mu^\mu = 0 \ , \tag{6.41a}$$

$$[B_\nu^\mu, B_\beta^\alpha] = \delta_\beta^\mu B_\nu^\alpha - \delta_\nu^\alpha B_\beta^\mu \ , \tag{6.41b}$$

which define the su(3) Lie algebra. Moreover, since A_μ and $A^\mu(\mu = 1, 2, 3)$ belong to 3 and $\bar{3}$ representations of the su(3), we must have

$$[B_\nu^\mu, A_\lambda] = \delta_\lambda^\mu A_\nu - \frac{1}{3} \delta_\nu^\mu A_\lambda \ , \tag{6.42a}$$

$$[B_\nu^\mu, A^\lambda] = -\delta_\nu^\lambda A^\mu + \frac{1}{3} \delta_\nu^\mu A^\lambda \ . \tag{6.42b}$$

Moreover, we know that

$$3 \otimes 3 = \bar{3} \oplus 6 \ ,$$
$$\bar{3} \otimes \bar{3} = 3 \oplus \bar{6} \ ,$$

from the representation theory of the su(3). These are translated to imply the validities of

$$[A_\mu, A_\nu] = -2 \sum_{\lambda=1}^{3} \epsilon_{\mu\nu\lambda} A^\lambda \ , \tag{6.43a}$$

$$[A^\mu, A^\nu] = 2 \sum_{\lambda=1}^{3} \epsilon^{\mu\nu\lambda} A_\lambda \ , \tag{6.43b}$$

where $\epsilon_{\mu\nu\lambda}$ and $\epsilon^{\mu\nu\lambda}$ are totally anti-symmetric Levi–Civita symbols in three dimension, normalized to $\epsilon^{123} = \epsilon_{123} = 1$. The factor ± 2 in the right sides of Eqs. (6.43) is simply for convenience. Finally, we note that

$$3 \otimes \bar{3} = 1 \oplus 8 \ ,$$

so that $[A_\nu, A^\mu]$ must be proportional to B_ν^μ. The multiplicative constant cannot, however, be arbitrary, since the Jacobi identities must be satisfied.

The correct answer is found to be

$$[A_v, A^\mu] = 3B_v^\mu \ . \tag{6.44}$$

The commutation relations (6.41), (6.42), (6.43), and (6.44) now define the 14-dimensional Lie algebra G_2.

The seven-dimensional module V of G_2 can be found first by reducing G_2 into su(3) with[51]

$$7 \to 7 = 1 \oplus 3 \oplus \bar{3} \ .$$

Therefore, the basis of V consists of su(3) vectors ϕ_μ and ϕ^μ ($\mu = 1, 2, 3$) belonging to 3 and $\bar{3}$, respectively, together with the su(3) singlet ξ. The actions of B_v^μ, A_μ, and A^μ on these basis vectors of V can easily be found to be

$$B_v^\mu \phi_\lambda = \delta_\lambda^\mu \phi_v - \frac{1}{3} \delta_v^\mu \phi_\lambda \ , \tag{6.45a}$$

$$B_v^\mu \phi^\lambda = -\delta_v^\lambda \phi^\mu + \frac{1}{3} \delta_v^\mu \phi^\lambda \ , \tag{6.45b}$$

$$B_v^\mu \xi = 0 \ , \tag{6.45c}$$

$$A_\mu \phi_v = \sum_{\lambda=1}^{3} \epsilon_{\mu v \lambda} \phi^\lambda \ , \tag{6.45d}$$

$$A^\mu \phi^v = -\sum_{\lambda=1}^{3} \epsilon^{\mu v \lambda} \phi_\lambda \ , \tag{6.45e}$$

$$A_\mu \phi^v = -2\delta_\mu^v \xi \ , \tag{6.45f}$$

$$A^\mu \phi_v = 2\delta_v^\mu \xi \ , \tag{6.45g}$$

$$A_\mu \xi = \phi_\mu \ , \tag{6.45h}$$

$$A^\mu \xi = -\phi^\mu \ . \tag{6.45i}$$

Let $x_j = \{\phi_{\mu j}, \phi_j^\mu, \xi_j\}$ ($j = 1, 2, 3$) be three generic coordinate elements of V with respect to the basis ϕ_μ, ϕ^μ, and ξ. Then the G_2-invariant symmetric bi-linear form is given by

$$< x_1 | x_2 > = 2\xi_1 \xi_2 + \sum_{\lambda=1}^{3} \{\phi_{\lambda 1} \phi_2^\lambda + \phi_1^\lambda \phi_{\lambda 2}\} \ , \tag{6.46}$$

while the completely anti-symmetric tri-linear form $\psi(x_1, x_2, x_3) \, \epsilon$ Hom

$([1^3] \to F)$ is found to be

$$\psi(x_1, x_2, x_3) = \sum_{\lambda=1}^{3} \{\phi_1^\lambda \phi_{\lambda 2} - \phi_2^\lambda \phi_{\lambda 1}\} \xi_3$$

$$+ \sum_{\lambda=1}^{3} \{\phi_2^\lambda \phi_{\lambda 3} - \phi_3^\lambda \phi_{\lambda 2}\} \xi_1$$

$$+ \sum_{\lambda=1}^{3} \{\phi_3^\lambda \phi_{\lambda 1} - \phi_1^\lambda \phi_{\lambda 3}\} \xi_2 \qquad (6.47)$$

$$+ \sum_{\mu,\nu,\lambda=1}^{3} \epsilon_{\lambda\mu\nu} \phi_1^\lambda \phi_2^\mu \phi_3^\nu$$

$$- \sum_{\mu,\nu,\lambda=1}^{3} \epsilon^{\lambda\mu\nu} \phi_{\lambda 1} \phi_{\mu 2} \phi_{\nu 3} \,.$$

Note that the coordinates $\phi_{\mu j}$, ϕ_j^μ, and ξ_j for $j = 1, 2, 3$ are ordinary numbers with $\xi_j = \; < \xi | x_j >$ and $\phi_{\mu j} = \; < \phi_\mu | x_j >$, for example. From Eqs. (6.46) and (6.47), we can compute the anti-symmetric product $x_j \cdot x_k$ as in the preceding sections with the multiplication table of

$$\phi_\mu \cdot \phi_\nu = \sum_{\lambda=1}^{3} \epsilon_{\mu\nu\lambda} \phi^\lambda \,, \qquad (6.48a)$$

$$\phi^\mu \cdot \phi^\nu = -\sum_{\lambda=1}^{3} \epsilon^{\mu\nu\lambda} \phi_\lambda \,, \qquad (6.48b)$$

$$\xi \cdot \phi_\mu = -\phi_\mu \cdot \xi = \frac{1}{\sqrt{2}} \, \phi_\mu \,, \qquad (6.48c)$$

$$\xi \cdot \phi^\mu = -\phi^\mu \cdot \xi = -\frac{1}{\sqrt{2}} \, \phi^\mu \,, \qquad (6.48d)$$

$$\phi_\mu \cdot \phi^\nu = -\phi^\nu \cdot \phi_\mu = \frac{1}{\sqrt{2}} \, \delta_\mu^\nu \xi \,. \qquad (6.48e)$$

In order to reduce this to the standard form of Eq. (6.8), we either identify

$$u_1 = \phi_1 \,, \quad u_2 = \phi_2 \,, \quad u_3 = -\phi_3 \,, \qquad (6.49a)$$

$$\bar{u}_1 = -\phi^1 \,, \quad \bar{u}_2 = -\phi^2 \,, \quad \bar{u}_3 = \phi^3 \,, \qquad (6.49b)$$

$$\psi_0 = \xi \,, \qquad (6.49c)$$

or, more symmetrically,

$$u_\mu = -\phi_\mu \,, \quad \bar{u}_\mu = \phi^\mu \,,$$
$$\psi_0 = \xi \,, \tag{6.50}$$

for all $\mu = 1, 2, 3$. The reason why we consider a rather asymmetrical assignment Eq. (6.49) will soon become apparent. Both then reproduce the multiplication table of Eq. (6.8). Moreover, if we extend V into $A = V^* = V \oplus Fe$, then we will obtain the multiplication table, Eq. (6.11), of the split Cayley algebra.

Remark 6.5. As the result of this section, together with the comment given in Remark 6.3, the derivation Lie algebra of the octonion algebra contains G_2 as its sub-algebras. The proof that the whole derivation algebra of the octonion is precisely G_2 will require further calculation.

(b) so(4) = su(2) ⊗ su(2) basis

If we choose the basis of G_2 next by considering its other maximal sub-Lie algebra[51] so(4), instead of su(3), then we can obtain another realization of the octonion algebra. The seven-dimensional representation of G_2 will then reduce according to

$$7 \to 7 = 3 + 4 \,,$$

under the reduction of $G_2 \to$ so(4), where four- and three-dimensional representations of the so(4) correspond to the 4-vector a_μ and self-dual tensor $f_{\mu\nu} = {}^* f_{\mu\nu}$ $(\mu, \nu = 1, 2, 3, 4)$ respectively. Then, after some calculations, we find the multiplication table of Eq. (1.40) which was used in Section 3.4 for the construction of the su(2)-instanton solution. The relationship between the so(4) basis, and the preceding su(3) basis of G_2 is as follows. Two pairs generated by B_2^1, B_1^2, and $B_1^1 - B_2^2$, and by A_3, A^3, and B_3^3, respectively, are the desired mutually commuting two su(2) sub-Lie algebras. The three-dimensional realization is spanned by ϕ^3, ϕ_3, and ξ, while ϕ_1, ϕ_2, ϕ^1, and ϕ^2 furnish the four-dimensional representation space of so(4).

(c) su(2) basis

G_2 also has the third maximal sub-Lie algebra su(2), where the seven-dimensional representation of G_2 remains unsplit as the seven-dimensional spin-3 realization of the su(2). Therefore, labeling G_2 on this basis, we can construct the octonion from the spin-3 module of the su(2), as we did in Section 6.1. This also explains why the spin-3 Clebsch–Gordan algebra of the su(2) leads to the octonion algebra. In this connection, we note that if L is the su(2), and if $V = [1]$ is the seven-dimensional spin-3 module of L, then we obtain

$$\text{Dim Hom } ([2,1] \rightarrow [1]) = 2 \, ,$$

so that the method described in the preceding section is *not* directly applicable for this case. Nevertheless, we have

$$\text{Dim Hom } ([1^2] \rightarrow [1]) = 1 \, ,$$

so that we can define the unique anti-symmetric product as in Section 6.1.

The maximal sub-Lie algebra su(2) ($\simeq$ SO(3)) of G_2 is now given by

$$
\begin{aligned}
J_+ &= \sqrt{6}\, A_1 + \sqrt{10}\, B_2^1 \, , \quad J_- = \sqrt{6}\, A^1 + \sqrt{10}\, B_1^2 \, , \\
J_3 &= 4\, B_1^1 + 5\, B_2^2 \, ,
\end{aligned}
\tag{6.51}
$$

which satisfies the familiar relation

$$[J_3, J_\pm] = \pm\, J_\pm \, , \quad [J_+, J_-] = 2\, J_3 \, . \tag{6.52}$$

Setting $\psi_m = |3, m >$ for $m = 0, \pm 1, \pm 2, \pm 3$, we can identify the ψ_ms in terms of ϕ_μ, ϕ^μ, and ξ, given by Eqs. (6.45), to be

$$
\begin{aligned}
\psi_3 &= \phi^3 \, , \quad \psi_2 = \phi_2 \, , \quad \psi_1 = \phi_1 \, , \quad \psi_0 = \sqrt{2}\, \xi \, , \\
\psi_{-1} &= -\phi^1 \, , \quad \psi_{-2} = \phi^2 \, , \quad \psi_{-3} = -\phi_3 \, .
\end{aligned}
\tag{6.53}
$$

We can verify the familiar formulae

$$
\begin{aligned}
J_3 \psi_m &= m\, \psi_m \, , \\
J_\pm \psi_m &= \sqrt{(3 \mp m)(4 \pm m)}\; \psi_{m \pm 1} \, ,
\end{aligned}
\tag{6.54}
$$

for the spin-3 module. Our present assignment is consistent with Eqs. (6.7) and (6.49). This is the reason why we considered the asymmetrical identification of Eq. (6.49), as well as the more symmetrical assignment of Eq. (6.50).

7

Algebra of physical observables

7.1 Jordan product

Flexible or Jordan algebras may be of some interest in physics. Let us consider the usual quantum mechanics. We then deal with various operators: Hamiltonian H, position operator q, momentum operator p, etc. They are not commutative but they are associative. In general, they correspond to linear operators in some Hilbert spaces, and observables in quantum mechanics are self-adjoint operators, so that their expectation values are real. Let x and y be observables, that is, $x^\dagger = x$, and $y^\dagger = y$. But then their product $z = xy$ may not necessarily be an observable since $z^\dagger = (xy)^\dagger = y^\dagger x^\dagger = yx$, which is not the same as $z = xy$ unless x and y commute. Therefore, if we wish to use only physically observable quantities, we may run into a problem. In order to avoid this difficulty, Jordan proposed[53] to use the Jordan product

$$z = x \cdot y = \frac{1}{2} (xy + yx) , \tag{7.1}$$

which is self-adjoint, $z^\dagger = z$ if $x^\dagger = x$ and $y^\dagger = y$. Therefore, the correct product is $x \cdot y$ and not the ordinary associative product. Note that the new product defines the real A^+-sub-algebra of an associative operator algebra A. Then, by the result of Eq. (2.21), we have

$$\begin{aligned} x \cdot y &= y \cdot x , \\ (x^2 \cdot y) \cdot x &= x^2 \cdot (y \cdot x) , \end{aligned} \tag{7.2}$$

so that the algebra of observables forms a real Jordan algebra.

Conversely, let us now temporarily forget the fact that we obtained Eq. (7.2) by Eq. (7.1) from an associative algebra, and let us adopt Eq. (7.2) as the definition of the algebra of observables. Since these are no longer associative, and the physically relevant case corresponds to an infinite-dimensional case, it is very difficult to analyze. However, if we assume that the Jordan algebra is of finite dimension, and if it is simple, then the famous theorem by Jordan, von Neuman, and Wigner[7,54] states the following result:

Theorem 9. (Jordan–von Neuman–Wigner). *If A is a finite-dimensional simple Jordan algebra, then A can be obtained in only one of the following two ways:*

<u>Case 1.</u> *The Jordan algebra can be obtained from some associative algebra as in Eq. (7.1). Such an algebra is known as a special Jordan algebra.*

<u>Case 2.</u> *Otherwise, the Jordan algebra is a 27-dimensional algebra which can be constructed as follows. Let X be a 3×3 matrix whose entries are octonionic numbers of the form*

$$ X = \begin{pmatrix} a_1 & x & y \\ \bar{x} & a_2 & z \\ \bar{y} & \bar{z} & a_3 \end{pmatrix}, \tag{7.3} $$

where a_1, a_2, $a_3 \in F$ are ordinary (real or complex) numbers and x, y, z are octonions with their conjugates $\bar{x}$, $\bar{y}$, $\bar{z}$. Define the product of two such matrices by

$$ X \cdot Y = \frac{1}{2} (XY + YX). \tag{7.4} $$

Then this defines the 27-dimensional exceptional Jordan algebra. Note that XY is a matrix product but it is <u>not</u> *associative, that is, $(XY)Z \neq X(YZ)$ in general, since its octonionic elements are not associative.*

Remark 7.1. The derivation algebra of the 27-dimensional exceptional Jordan algebra is the exceptional Lie algebra F_4. Conversely, the exceptional Jordan algebra can be[50] realized as the Clebsch–Gordan algebra of the 26-dimensional irreducible representation of F_4, together with the unit element, just as for the octonionic case studied in the preceding section.

Remark 7.2. Since physically relevant observables in quantum mechanics (QM) will be infinite dimensional, the theorem by Jordan–von Neuman–Wigner is not directly applicable to QM. More recently, Zelmanov[55] has proved, however, that there is *no* infinite-dimensional simple exceptional Jordan algebra. This theorem appears to dash the original hope of

Jordan's constructing a new type of QM. However, it may be noted that even in standard QM, physical observables are unbounded operators, so that their products are not always well defined except in a dense domain. Therefore, it may be conceivable that the theorem by Zelmanov may be circumvented if the Jordan product is defined, not universally, but only in a dense subset of some topological space. Also, the physical significance of the simpleness condition is not so obvious.

7.2 Heisenberg approach

Let us study QM now from a slightly different viewpoint. In the usual associative QM, we obtain the time evolution of any operator (say x) from the Heisenberg equation of motion for a Hamiltonian H:

$$\frac{d}{dt} x = i[H, x] , \tag{7.5}$$

where $[x, y] = xy - yx$ is the commutator. Again, forget the fact that the underlying operator algebra is associative, but assume that the Heisenberg equation of motion is nevertheless valid. We may ask the question: What is the most general non-associative algebra A which is compatible with Eq. (7.5)? First, consider the consistency condition

$$\frac{d}{dt} (xy) = x \frac{d}{dt} y + \frac{dx}{dt} y , \tag{7.6}$$

which leads to

$$[H, xy] = x[H, y] + [H, x]y . \tag{7.7}$$

The validity of Eq. (7.7) is not obvious for a general algebra. However, we note the following theorem.

Theorem 10. (**Myung**[22]). *A necessary and sufficient condition that we have*

$$[z, xy] = [z, x]y + x[z, y] \tag{7.8}$$

for any x, y, $z \in A$ is that A be flexible and Lie-admissible, that is, that we have

$$
\begin{align}
&\text{(i)} \quad (x, y, z) = -(z, y, x) , \\
&\text{(ii)} \quad [[x, y], z] + [[z, x], y] + [[y, z], x] = 0 .
\end{align} \tag{7.9}
$$

Because of Theorem 10, the consistency condition Eq. (7.7) is automatically satisfied if A is a flexible Lie-admissible algebra. Another advantage of Eq. (7.8) is that it will also assure the self-consistency of quantization.

In other words, we must verify that the Heisenberg equation of motion will reproduce the Lagrange equation of motion correctly. Instead of the customary associative law, it is really sufficient,[56] in general, to have only Eq. (7.8). For these reasons, a study of such an algebra may be of some interest. There are many discussions of some simple cases (see references 22, and 56–59).

One interesting example of a flexible Lie-admissible algebra is the so-called mutation algebra by Santilli.[59] Let A be an associative algebra over a field F. For any arbitrary, but fixed, element $p \in A$ and for any $\lambda \in F$, we may define a deformed product $x * y$ for x, $y \in A$ in the same vector space A by

$$x * y = xpy + \lambda ypx .$$

The resulting deformed algebra A^* can be readily verified to be flexible and Lie-admissible.

In the above discussion, we demanded the validity of Eq. (7.7). However, we can relax this as follows. Just as in the original reasoning of Jordan, xy may not be an observable. Hence, we may restrict the validity of the Heisenberg equation to physical observable quantities only. Then we may argue that we should apply it to a Jordan product

$$x \cdot y = \frac{1}{2} (xy + yx)$$

rather than to xy (or yx). The consistency condition now becomes

$$\frac{d}{dt} (x \cdot y) = \frac{dx}{dt} \cdot y + x \cdot \frac{dy}{dt} ,$$

which gives a weaker condition:

$$[H, x \cdot y] = [H, x] \cdot y + x \cdot [H, y] , \qquad (7.10)$$

in comparison to Eq. (7.7). Note that the validity of Eq. (7.7) implies that of Eq. (7.10) but not, in general, conversely. We note the following theorem.

Theorem 11. *A necessary and sufficient condition to have*

$$[z, x \cdot y] = [z, x] \cdot y + x \cdot [z, y]$$

for three elements x, y, z $\in$ A is that A be flexible.

Proof. Note the identity

$$[z, x] \cdot y + x \cdot [z, y] - [z, x \cdot y]$$
$$= \frac{1}{2} \{ (x, y, z) + (y, x, z) + (z, x, y) + (z, y, x) - (x, z, y) - (y, z, x) \},$$
$$(7.11)$$

from which we can readily prove[22] the theorem, where $(x, y, z) = (xy)z - x(yz)$.

Remark 7.3. Theorem 11 requires only the flexibility condition but *not* Lie-admissibility as in Theorem 10. If A is flexible, then Eq. (7.10) is automatically valid. Therefore, we must at least impose the flexibility condition for the algebra of observables.

Remark 7.4. Within the framework of the algebra of observables, the Heisenberg equation of motion may be regarded as more fundamental than the Schrödinger equation. As a matter of fact, the Schrödinger equation is rather difficult to define, although one way to do so[56] is to consider faithful representations of the associated Lie algebra A^-, if A is Lie-admissible.

Remark 7.5. In regard Jordan's approach, the Heisenberg equation of motion may be interpreted in the following ways. If we were to interpret the commutator in Jordan form $[H, x] = H \cdot x - x \cdot H$, then it would be identically zero, so that all observables would be time-independent, corresponding to a kind of Schrödinger representation. Alternatively, we may suppose that A is not commutative but Jordan-admissible, with $[H, x] = Hx - xH$ being non-zero. Then we can utilize Theorem 11, so that the likeliest candidate for A will be a flexible Jordan-admissible algebra, that is, A will be the so-called non-commutative Jordan algebra.[7]

Remark 7.6. In classical mechanics, physical observables are real functions of coordinates and momenta of particles, while the Heisenberg equation of motion is replaced by Hamilton's equation. It has been observed by Kantor[60] that there is a one-to-one correspondence between the classical Poisson bracket and a class of Jordan-super algebra. Therefore, both quantum and classical mechanics may be formulated in terms of a general Jordan-super algebra. However, its possible physical significance is not yet clear.

7.3 Non-linear Schrödinger equation

Here, we will briefly discuss observables from the viewpoint of Schröd-inger wave functions. For simplicity, we consider only the case of a finite-dimensional Hilbert space V with $N = \dim V < \infty$. Let $e_1, e_2, \ldots, e_N$ be a basis of V so that any vector $\psi \in V$ may be expressed as

$$\psi = \sum_{j=1}^{N} \psi_j e_j , \quad \psi_j \in F .$$ (7.12)

For simplicity, we will write this as

$$\psi = \{\psi_1, \psi_2, \ldots, \psi_N\} ,$$ (7.13)

with its complex conjugate

$$\psi^+ = \{\psi_1^+, \psi_2^+, \ldots, \psi_N^+\} .$$ (7.14)

Following Weinberg,[61] we define an observable b to be a real *non-linear* function $b(\psi, \psi^+)$. We note that if b has a bi-linear structure

$$b(\psi, \psi^+) = \sum_{j,k=1}^{N} \psi_j^+ B_{jk} \psi_k$$ (7.15)

for some constants B_{jk}, then we may identify b with the $N \times N$ matrix B defined by $<j|B|k> = B_{jk}$. This corresponds to standard quantum mechanics. Moreover, the reality of b is equivalent to the self-adjointness of the matrix B. If we wished, we could impose the additional condition[61] that both ψ and $Z\psi = \{Z\psi_1, Z\psi_2, \ldots, Z\psi_N\}$, for any complex number Z, should represent the same state as in standard quantum mechanics. This implies that the function b must be a homogeneous function of degree 1 for both ψ and ψ^+, that is,

$$\sum_{j=1}^{N} \psi_j \frac{\partial b}{\partial \psi_j} = \sum_{j=1}^{N} \psi_j^+ \frac{\partial b}{\partial \psi_j^+} = b ,$$ (7.16)

where we will often drop the ψ and ψ^+ dependence of $b(\psi, \psi^+)$ for simplicity. However, since the validity of Eq. (7.16) is not essential for the following discussion, we will not assume this until later.

A set of all such functions of ψ and ψ^+, but without the reality condition, forms a complex algebra as follows. First, we define their sum $a + b$ by

$$(a + b)(\psi, \psi^+) = a(\psi, \psi^+) + b(\psi, \psi^+) .$$ (7.17)

Next, following Weinberg[61], we define their product by

$$a * b = \sum_{j=1}^{N} \frac{\partial a}{\partial \psi_j} \frac{\partial b}{\partial \psi_j^+} . \tag{7.18}$$

If a and b are bi-linear as in Eq. (7.15), with the corresponding matrices A and B, respectively, then $a * b$ corresponds to the matrix product AB, so that the product Eq. (7.18) is a generalization of the standard operator product in QM. However, it is no longer associative in general. Nevertheless, it is Lie-admissible since the bracket

$$[a, b] \equiv a * b - b * a = \sum_{j=1}^{N} \left(\frac{\partial a}{\partial \psi_j} \frac{\partial b}{\partial \psi_j^+} - \frac{\partial a}{\partial \psi_j^+} \frac{\partial b}{\partial \psi_j} \right) \tag{7.19}$$

is essentially the classical Poisson bracket. Unfortunately, it is *not* flexible, since we calculate

$$(a * b) * a - a * (b * a)$$

$$= \sum_{j,k=1}^{N} \left\{ \frac{\partial^2 a}{\partial \psi_j \partial \psi_k} \frac{\partial b}{\partial \psi_k^+} \frac{\partial a}{\partial \psi_j^+} - \frac{\partial^2 a}{\partial \psi_j^+ \partial \psi_k^+} \frac{\partial b}{\partial \psi_k} \frac{\partial a}{\partial \psi_j} \right\} , \tag{7.20}$$

which is not zero. This now causes the following serious problem if, as in reference 61, we assume the Heisenberg equation of motion

$$\frac{\mathrm{d}}{\mathrm{d}t} a = -\mathrm{i}[a, h] = -\mathrm{i} \sum_{j=1}^{N} \left\{ \frac{\partial a}{\partial \psi_j} \frac{\partial h}{\partial \psi_j^+} - \frac{\partial a}{\partial \psi_j^+} \frac{\partial h}{\partial \psi_j} \right\} \tag{7.21}$$

for a Hamiltonian observable h. Since the present algebra is *not* flexible, there is no guarantee of the validity of the consistency condition Eq. (7.7) and/or Eq. (7.10). Actually, by direct calculation we can find

$$a * [b, h] + [a, h] * b - [a * b, h] = a \cdot [b, h] + [a, h] \cdot b - [a \cdot b, h]$$

$$= \sum_{j,k=1}^{N} \left\{ \frac{\partial a}{\partial \psi_j} \frac{\partial b}{\partial \psi_k} \frac{\partial^2 h}{\partial \psi_j^+ \partial \psi_k^+} - \frac{\partial a}{\partial \psi_j^+} \frac{\partial b}{\partial \psi_k^+} \frac{\partial^2 h}{\partial \psi_j \partial \psi_k} \right\} . \tag{7.22}$$

Here we have set

$$a \cdot b = \frac{1}{2} \{ a * b + b * a \}$$

and the validity of the first relation in Eq. (7.22) is due to the Lie-admissible property of the algebra. Therefore, unless we have

$$\frac{\partial^2 h}{\partial \psi_j \partial \psi_k} = \frac{\partial^2 h}{\partial \psi_j^+ \partial \psi_k^+} = 0 , \tag{7.23}$$

the consistency condition, Eq. (7.7) or (7.10), for $h = H$ is violated. Note that Eq. (7.23) forces h to be a quadratic form:

$$h = \sum_{j,k=1}^{N} \psi_j^+ H_{jk} \psi_k + \sum_{j=1}^{N} \left(\psi_j^+ D_j + D_j^+ \psi_j \right) + B_0 \qquad (7.24)$$

for some constants H_{jk}, D_j, and B_0. If we ignore the D_j and B_0 terms on physical grounds, then this gives only the standard Hamiltonian observable, with no non-linear effect.

Now, in addition, let us assume the CP^{N-1} constraint equation, Eq. (7.16), for all observables. In this case, we can make the right side of Eq. (7.22) vanish identically, provided that h satisfies

$$\frac{\partial^2 h}{\partial \psi_j \partial \psi_k} = \psi_j^+ \psi_k^+ g \,, \qquad \frac{\partial^2 h}{\partial \psi_j^+ \partial \psi_k^+} = \psi_j \psi_k g \,, \qquad (7.25)$$

for some real function $g = g(\psi, \psi^+)$. However, differentiating Eq. (7.16) we obtain

$$\sum_{j=1}^{N} \psi_j \frac{\partial^2 b}{\partial \psi_j \partial \psi_k} = 0 = \sum_{j=1}^{N} \psi_j^+ \frac{\partial^2 b}{\partial \psi_j^+ \partial \psi_k^+} \,. \qquad (7.26)$$

Therefore, for the choice $b = h$, Eqs. (7.25) and (7.26) imply that $g = 0$, reproducing condition (7.23), and hence Eq. (7.24), again.

In conclusion, it would appear that the only choice left to save the interesting non-linear aspect of the theory (see however Remark 7.9 below) is to abandon the Leibnitz rule

$$\frac{d}{dt}(a * b) = \frac{da}{dt} * b + a * \frac{db}{dt} \,. \qquad (7.27)$$

If this is so, then the product $a * b$ inherently violates time-translation invariance.

Remark 7.7. As we see from Eq. (7.20) with $b = a$, the present algebra is not third-order power associative, that is, it does not satisfy the third-order power associative law

$$(a * a) * a = a * (a * a) \,, \qquad (7.28)$$

as has been observed in reference 61. In contrast, any flexible algebra, such as an octonion, pseudo-octonion or Jordan algebra, is automatically third-order power-associative.

Remark 7.8. In ordinary quantum mechanics, the Heisenberg picture is equivalent to the Schrödinger picture. However, the same will not necessarily be so for possible non-associative quantum mechanics. The idea of a Hilbert space realization of non-associative algebras is not an easy concept, in contrast with associative cases, except for those of Jordan algebras as well as Lie-admissible algebras, although we will not go into detail here. However, for some special non-associative algebras, we can define the corresponding realization in a Hilbert space. For example, one can consult reference 62 in connection with the so-called nuclear boson-representation of fermion Fock spaces.

Remark 7.9. We can overcome the difficulty encountered in this section if we use the associative Moyal product instead of Eq. (7.18) for $a*b$. In that case, Eq. (7.19) is replaced by the Moyal bracket. The resulting theory, however, corresponds to the second quantization. For the definition of the Moyal product and bracket, see references 62a.

8

Triple products and ternary systems

8.1 Introduction

The bi-linear products in a vector space V can be identified by a linear mapping:

$$V \otimes V \to V . \tag{8.1}$$

It is thus natural to consider more general linear mappings such as

$$f : V \otimes V \otimes V \to V , \tag{8.2}$$

which introduces the notion of a triple product in V. For any x, y, $z \in V$, we assign an element $w \in V$, which is linear in each of x, y, and z, and we write $w = [x, y, z]$, or $< x, y, z >$, or even $w = xyz$ as a juxtaposition whenever there is no confusion with possible double products of some bi-linear product, that is, $(xy)z$ or $x(yz)$ in V. The vector space V possessing a triple product will be called a triple (or ternary) system.

Interesting cases are obtained, however, when V is a module of a Lie algebra L and the product is covariant under the action of L, that is,

$$b[x, y, z] = [bx, y, z] + [x, by, z] + [x, y, bz] , \tag{8.3}$$

or

$$f \circ b = b \circ f , \tag{8.4}$$

for the mapping f given by Eq. (8.2), just as in Eq. (6.17), since $b \in L$ acts on $V \otimes V \otimes V$ according to

$$b(x \otimes y \otimes z) = (bx) \otimes y \otimes z + x \otimes (by) \otimes z + x \otimes y \otimes (bz) .$$

The necessary condition for the existence of a non-trivial L-covariant triple product in V is to have

$$\text{Dim Hom } (V \otimes V \otimes V \to V) \geq 1 .$$

Note that since we could have Dim Hom $(V \otimes V \to V) = 0$, the space V may not possess any (L-covariant) bi-linear product, even though it has a (L-covariant) triple product. Such a case occurs,[50] for example, for the 56-dimensional irreducible module of the Lie algebra E_7.

Any linear mapping $D : V \to V$ satisfying

$$D[x, y, z] = [Dx, y, z] + [x, Dy, z] + [x, y, Dz] \qquad (8.5)$$

is called a derivation of the triple system. In particular, any $b \in L$ is a derivation of V. Let D_1 and D_2 be two derivations. It is easy to see that the commutator $D = [D_1, D_2]$ is also a derivation. Therefore, a set consisting of all derivations of a triple system V forms a Lie algebra, called the derivation algebra of V. In particular, the original Lie algebra L is its sub-Lie algebra.

Before going into further details, it may be worth while considering a simple example. Let V be a four-dimensional real or complex vector space with basis vectors e_1, e_2, e_3, and e_4 satisfying

$$< e_\mu | e_\nu > = \delta_{\mu\nu} , \quad (\mu, \nu = 1, 2, 3, 4) , \qquad (8.6)$$

which is clearly invariant under the action of any SO(4) group and hence of $L = $ so(4). Now we can define the unique L-covariant totally anti-symmetric triple product $[x, y, z]$ by

$$[e_\mu, e_\nu, e_\lambda] = \sum_{\alpha=1}^{4} \epsilon_{\mu\nu\lambda\alpha} e_\alpha , \qquad (8.7)$$

where $\epsilon_{\mu\nu\lambda\alpha}$ is the totally anti-symmetric Levi–Civita symbol with $\epsilon_{1234} = 1$, as before. The uniqueness of the triple product follows easily from

$$\text{Dim Hom} ((V \otimes V \otimes V)_A \to V) = 1 , \qquad (8.8a)$$

or, equivalently,

$$\text{Dim Hom} ([1^3] \to [1]) = 1 , \qquad (8.8b)$$

in the notation of the preceding section.

We can, moreover, calculate the following equation by Eq. (8.7):

$$
\begin{aligned}
\left[e_\alpha, e_\beta, \left[e_\mu, e_\nu, e_\lambda\right]\right] &= \sum_{\gamma=1}^{4} \epsilon_{\mu\nu\lambda\gamma} \left[e_\alpha, e_\beta, e_\gamma\right] \\
&= \sum_{\gamma=1}^{4} \epsilon_{\mu\nu\lambda\gamma} \sum_{\theta=1}^{4} \epsilon_{\alpha\beta\gamma\theta} e_\theta \\
&= -\left(\delta_{\mu\alpha}\delta_{\nu\beta} - \delta_{\nu\alpha}\delta_{\mu\beta}\right)e_\lambda - \left(\delta_{\nu\alpha}\delta_{\lambda\beta} - \delta_{\lambda\alpha}\delta_{\nu\beta}\right)e_\mu \\
&\quad - \left(\delta_{\lambda\alpha}\delta_{\mu\beta} - \delta_{\mu\alpha}\delta_{\lambda\beta}\right)e_\nu \; ,
\end{aligned}
$$

which can be rewritten as a triple product relation:

$$
\begin{aligned}
[u, v, [x, y, z]] = &- (<u|x><v|y> \; - \; <u|y><v|x>)z \\
&- (<u|y><v|z> \; - \; <u|z><v|y>)x \qquad (8.9) \\
&- (<u|z><v|x> \; - \; <u|x><v|z>)y
\end{aligned}
$$

for any u, v, x, y, $z \, \epsilon \, V$. Conversely, we can prove that any triple system satisfying Eq. (8.9) for a completely anti-symmetric non-trivial product $[x, y, z]$ must be four dimensional, that is, Dim $V = 4$. Further, if we choose a suitable basis, it will satisfy Eq. (8.7). We call such a triple system a quaternionic triple system, as we will see in Section 8.4.

8.2 Alternative triple systems and Hurwitz algebras

Let V be a vector space with $N = $ Dim V over a field F. We suppose that it possesses a totally anti-symmetric triple product $[x, y, z]$, although we do not yet specify the underlying Lie algebra L. Further, we assume, first, the existence of the function

$$
B(x, y|z) \equiv B_z(x, y) \, \epsilon \, F \; , \qquad (8.10)
$$

which is bi-linear in x and y but quadratic in $z \, \epsilon \, V$, satisfying the following condition:

$$
B_z(x, x) = B_x(z, z) \; . \qquad (8.11)
$$

Second, we assume the validity of the triple product relation

$$
[x, [x, y, z], z] = B_x(y, z)z + B_z(y, x)x - B_z(x, x)y \qquad (8.12)
$$

for any x, y, $z \, \epsilon \, V$. Note that Eq. (8.11) is necessary for the consistency of Eq. (8.12) when we interchange $x \leftrightarrow z$. A completely anti-symmetric triple product $[x, y, z]$ satisfying Eq. (8.12), is called an alternative triple product in reference 63, under the additional condition

$$B_z(y, x) = B_z(x, y) , \qquad (8.13)$$

which we will *not* yet assume, however.

Next, let $\phi(x) \in F$ be a linear functional (or linear form) in V and let $e \in V$ be any arbitrary, but fixed, element of V satisfying

$$\phi(e) = 1 . \qquad (8.14)$$

We introduce a bi-linear form $< x|y >$ and a bi-linear product xy in V by

$$< x|y > = \phi(x)\phi(y) + \lambda^2 B_e(y, x) \qquad (8.15a)$$

$$xy = \lambda\{[x, y, e] - \phi([x, y, e])e\} + \phi(x)y + \phi(y)x - < x|y > e , \qquad (8.15b)$$

for a constant λ. We will now prove the following theorem.

Theorem 12. *Let V be an alternative triple system which may not, however, satisfy the symmetry condition Eq. (8.13). We then obtain*

(i) $\phi(x) = < x|e > = < e|x > ,$ \qquad (8.16a)

(ii) $xe = ex = x ,$ \qquad (8.16b)

(iii) $xx - 2\phi(x)x + < x|x > e = 0 ,$ \qquad (8.16c)

(iv) $x(xy) = (xx)y ,$ \qquad (8.16d)

(v) $< x|xy > = \phi(y) < x|x > ,$ \qquad (8.16e)

(vi) $< xy|xy > = < x|x >< y|y > + \phi(y)\{< xy|x > - < x|xy >\} .$ \qquad (8.16f)

In particular, if $B_z(x, y)$ satisfies the symmetry condition of Eq. (8.13), then the bi-linear algebra defined by the product xy is an alternative composition algebra. Therefore, it will define a Hurwitz algebra whenever $< x|y > = < y|x >$ is non-degenerate.

Remark 8.1. Any algebra satisfying Eq. (8.16d) is termed left-alternative. The present theorem generalizes the one given in ref. 63.

Proof To prove Theorem 12, we first observe the following identities:

$$B_z(x, z) = B_z(z, x) = 0 , \qquad (8.17a)$$

$$B_z([x, y, z], x) = 0 . \qquad (8.17b)$$

First, setting $y = x$ in Eq. (8.12) and using $[x, x, z] = 0$, we obtain

$B_x(x, z) = 0$ and hence $B_z(z, x) = 0$ by interchanging $x \leftrightarrow z$. Second, when we set $x = z$ in Eq. (8.12) and note that $[z, y, z] = 0$, we find, similarly,

$$2\, B_z(y, z)z - B_z(z, z)y = 0 \,,$$

which leads to $B_z(z, z) = 0$ by further setting $y = z$, and hence to $B_z(y, z) = 0$. Substituting y by x, this proves the validity of Eq. (8.17a). In order to prove Eq. (8.17b), we replace y in Eq. (8.12) by

$$w = [x, y, z] \tag{8.18a}$$

to obtain

$$[x, [x, w, z], z] = B_x(w, z)z + B_z(w, x)x - B_z(x, x)w \,. \tag{8.18b}$$

On the other hand, we calculate

$$[x, w, z] = [x, [x, y, z], z]$$
$$= B_x(y, z)z + B_z(y, x)x - B_z(x, x)y \,,$$

so that $[x, [x, w, z], z] = -B_z(x, x)[x, y, z] = -B_z(x, x)w$. Comparing this with Eq. (8.18b), we obtain

$$B_x(w, z)z + B_z(w, x)x = 0 \,.$$

If x and z are linearly independent, this immediately gives the desired relation $B_z(w, x) = 0$. If x and z are linearly dependent, we also have $B_z(w, x) = 0$ trivially, since then $w = [x, y, z] = 0$. This completes the proof of Eq. (8.17b).

When we set $z = e$, Eqs. (8.17) give

$$B_e(x, e) = B_e(e, x) = 0 \,, \tag{8.19a}$$
$$B_e([x, y, e], x) = 0 \,. \tag{8.19b}$$

Then, Eqs. (8.16a), (8.16b), and (8.16c) are immediate consequences of Eq. (8.19a), together with Eqs. (8.14) and (8.15).

Next, in order to prove Eq. (8.16d), we first set

$$x \wedge y = [x, y, e] - \phi([x, y, e])e \tag{8.20}$$

for simplicity. The $\wedge$ operation satisfies

$$x \wedge y = -y \wedge x \,, \quad x \wedge e = 0 \,, \tag{8.21a}$$
$$\phi(x \wedge y) = 0 \,, \tag{8.21b}$$

as well as

$$< x | x \wedge y > = 0 \,, \tag{8.21c}$$

We note here that Eq. (8.21c) is a consequence of Eqs. (8.20) and (8.19b). Moreover, we can show by calculation that

$$x \wedge (x \wedge y) = [x, [x, y, e], e] - \phi([x, [x, y, e], e])e$$
$$= B_e(y, x)x - B_e(x, x)y + \{B_e(x, x)\phi(y) - B_e(y, x)\phi(x)\}e . \tag{8.22}$$

Rewriting Eq. (8.15b) as

$$xy = \lambda x \wedge y + \phi(x)y + \phi(y)x - < x|y > e , \tag{8.23}$$

we then obtain the desired relation Eq. (8.16d), since we obtain

$$(xx)y - x(xy) = \lambda < x|x \wedge y > = 0 .$$

Also, Eq. (8.23) implies

$$< x|xy > = \lambda < x|x \wedge y > + \phi(x) < x|y > + \phi(y) < x|x >$$
$$- < x|y >< x|e > ,$$

which gives Eq. (8.16e) by Eq. (8.21c).

Finally, after linearizing Eq. (8.16e) by letting $x \to x \pm z$, we can rewrite it as

$$< x|zy > + < z|xy > = \phi(y)\{< x|z > + < z|x >\} .$$

By changing z into xy, and utilizing Eqs. (8.16c) and (8.16d), we obtain Eq. (8.16f) after some calculations. This completes the proof of Theorem 12. ∎

Remark 8.2. There are other identities involving $B_z(y, x)$. See reference 63 for the details.

Remark 8.3. For all of our applications to be given below, we assume that $B_z(x, y)$ always satisfies the symmetry condition Eq. (8.13), so that we also have $< x|y > = < y|x >$. Moreover, $< x|y >$ is non-degenerate, and hence the resulting algebra is a Hurwitz algebra. Also, apart from the example given in Section 8.3(c), $B_z(x, y)$ has one of the following two forms, either

$$B_z(x, y) = (x|y)(z|z) - (x|z)(y|z) \tag{8.24}$$

or

$$B_z(x, y) = \psi(x)\psi(y)(z|z) + \psi(z)\psi(z)(x|y)$$
$$- \psi(x)\psi(z)(y|z) - \psi(y)\psi(z)(x|z) , \tag{8.25}$$

where $(x|y) = (y|x)$ is a symmetric bi-linear non-degenerate form in V

and where $\psi(x)$ is a linear functional in V. For the first form, given by
Eq. (8.24), it is convenient to choose $\lambda = 1$ and $e \in V$ satisfying $(e|e) = 1$,
with $\phi(x) = (x|e)$. The bi-linear form $< x|y >$ given by Eq. (8.15a)
coincides with $(x|y)$, that is, we now have

$$< x|y > = (x|y) , \qquad (8.26)$$

which, in particular, is non-degenerate. For the second form, given by
Eq. (8.25), we choose $\phi(x)$ to be $\psi(x)$, with $\lambda = 1$. Then we again find
Eq. (8.26) to be valid. Thus if $B_z(x, y)$ is given by either Eq. (8.24) or
(8.25), we can construct a Hurwitz algebra by Theorem 12. Conversely,
we can prove the following theorem.

Theorem 13. *Let A be a bi-linear algebra with an anti-symmetric bi-linear
product* $[x, y]$. *Let* $\psi(x)$ *be a linear functional of A, satisfying*

$$\psi([x, y]) = 0$$

identically and let $K(y, x)$ *and* $F(y, x)$ *be bi-linear forms, which need not
be symmetric. Suppose that we have*

$$[x, [x, y]] = K(y, x)x - K(x, x)y + \{\psi(y)F(x, x) - \psi(x)F(y, x)\}f \quad (8.27)$$

for a fixed element $f \in A$. *Then the triple linear form defined by*

$$[x, y, z] = \frac{1}{2} \{\psi(x)[y, z] + \psi(y)[z, x] + \psi(z)[x, y]\} \qquad (8.28a)$$

is an alternative triple system satisfying Eq. (8.12), with

$$\begin{aligned} 4\,B_z(y, x) =&\, \psi(z)\psi(z)K(y, x) + \psi(x)\psi(y)K(z, z) \\ &- \psi(z)\psi(y)K(z, x) - \psi(x)\psi(z)K(y, z) . \end{aligned} \qquad (8.28b)$$

Remark 8.4. Any Hurwitz algebra satisfies the condition of Theorem 13,
with $\psi(x) = < x|f >$, where f is the unit element. In that case we have
$K(y, x) = F(y, x) = 4 < y|x >$. We can then construct another Hurwitz
algebra from Theorem 12. However, we will not go into details here.

8.3 Zorn's vector matrix octonion algebra and the Cayley–Dickson process

As applications of the result given in Section 8.2, we will now discuss two
constructions of octonion algebras associated with Zorn and Cayley–
Dickson (see reference 7).

(a) Zorn's vector matrix octonion algebra

Let $\mathbf{u}$ and $\mathbf{v}$ be two generic vectors in a three-dimensional vector space with the usual inner product denoted by

$$(\mathbf{u}, \mathbf{v}) = (\mathbf{v}, \mathbf{u}) . \tag{8.29}$$

We now consider three sets of vectors $\mathbf{u}_j$ and $\mathbf{v}_j$, as well as of constants α_j, $\beta_j \,\epsilon\, F$ for $j = 1, 2, 3$. We now set

$$x_j = \begin{pmatrix} \alpha_j, & \mathbf{u}_j \\ \mathbf{v}_j, & \beta_j \end{pmatrix} \qquad (j = 1, 2, 3) \tag{8.30}$$

and define a triple-linear alternative product by

$$[x_1, x_2, x_3] = \begin{pmatrix} \alpha, & \mathbf{u} \\ \mathbf{v}, & \beta \end{pmatrix} , \tag{8.31}$$

with two possible sign choices of

$$\beta = -\alpha = \pm \frac{1}{4} \sum_{j,k,\ell=1}^{3} \epsilon_{jk\ell}(\alpha_j + \beta_j)(\mathbf{u}_k, \mathbf{v}_\ell) , \tag{8.32a}$$

$$\mathbf{u} = -\frac{1}{2} \sum_{j,k,\ell=1}^{3} \epsilon_{jk\ell} \left\{ \alpha_j \beta_k \mathbf{u}_\ell \mp \frac{1}{2} (\alpha_j + \beta_j) \, \mathbf{v}_k \times \mathbf{v}_\ell \right\} , \tag{8.32b}$$

$$\mathbf{v} = \frac{1}{2} \sum_{j,k,\ell=1}^{3} \epsilon_{jk\ell} \left\{ \alpha_j \beta_k \mathbf{v}_\ell + \frac{1}{2} (\alpha_j + \beta_j) \mathbf{u}_k \times \mathbf{u}_\ell \right\} . \tag{8.32c}$$

Where $\mathbf{u} \times \mathbf{v}$ represents the standard vector product of two three-dimensional vectors $\mathbf{u}$ and $\mathbf{v}$. When we use the formula

$$\mathbf{u} \times (\mathbf{v} \times \mathbf{w}) = (\mathbf{u}, \mathbf{w})\mathbf{v} - (\mathbf{u}, \mathbf{v})\mathbf{w} ,$$

we can prove the validity of Eq. (8.12), that is,

$$[x_1, [x_1, x_2, x_3], x_3] = B_{x_3}(x_2, x_1) x_1 + B_{x_1}(x_2, x_3) x_3 - B_{x_3}(x_1, x_1) x_2 , \tag{8.33}$$

where $B_z(x, y)$ is given by Eq. (8.25), with

$$(x_j|x_k) = \frac{1}{2} (\alpha_j \beta_k + \alpha_k \beta_j) \pm \frac{1}{2} \{ (\mathbf{u}_j, \mathbf{v}_k) + (\mathbf{u}_k, \mathbf{v}_j) \} , \tag{8.34a}$$

$$\psi(x_j) = \frac{1}{2} (\alpha_j + \beta_j) . \tag{8.34b}$$

Note that for the choice of $\lambda = 1$ and

$$e = \begin{pmatrix} 1 & 0 \\ 0 & 1 \end{pmatrix} \tag{8.35}$$

for the unit element, we have $\psi(x_j) = <e|x_j>$ and $<x|y> = (x|y)$. We can compute the product xy from Eq. (8.15b). However, in order to avoid possible confusion with the usual matrix product, we will use the notation $x \cdot y$ for the product. The result is

$$\begin{pmatrix} \alpha_1 , & \mathbf{u}_1 \\ \mathbf{v}_1 , & \beta_1 \end{pmatrix} \cdot \begin{pmatrix} \alpha_2 , & \mathbf{u}_2 \\ \mathbf{v}_2 , & \beta_2 \end{pmatrix}$$

$$= \begin{pmatrix} \alpha_1\alpha_2 \mp (\mathbf{u}_1,\mathbf{v}_2) , & \beta_2\mathbf{u}_1 + \alpha_1\mathbf{u}_2 \pm \mathbf{v}_1 \times \mathbf{v}_2 \\ \alpha_2\mathbf{v}_1 + \beta_1\mathbf{v}_2 + \mathbf{u}_1 \times \mathbf{u}_2 , & \beta_1\beta_2 \mp (\mathbf{u}_2,\mathbf{v}_1) \end{pmatrix} . \tag{8.36}$$

This is precisely the standard form[7] of Zorn's vector matrix product, where we will use familiar expressions from various authors.[5,7,64,65] If we adopt a suitable basis for the e_μs, then Zorn's vector matrix algebra leads to the multiplication table of the split Cayley algebra.

(b) Cayley–Dickson process

Let B be an *associative* algebra with unit element 1, satisfying the quadratic equation

$$aa - 2 < a|1 > a + \ <a|a> 1 = 0 \tag{8.37}$$

for any $a \ \epsilon \ B$, where $< a|b >$ is a non-degenerate symmetric bi-linear form in B. Consider a vector space V given by

$$V = B \oplus B , \tag{8.38}$$

so that

$$\mathrm{Dim}\ V = 2\ \mathrm{Dim}\ B . \tag{8.39}$$

For any three

$$x_j = a_j \oplus b_j \ \epsilon \ V , \quad (j = 1,2,3) , \tag{8.40}$$

we define

$$[x_1, x_2, x_3] = a \oplus b \tag{8.41a}$$

by

$$a = \frac{1}{2} \sum_{j,k,\ell=1}^{3} \epsilon_{jk\ell} < a_j | 1 > \{a_k a_\ell - \mu b_k \bar{b}_\ell\} , \tag{8.41b}$$

$$b = \frac{1}{2} \sum_{j,k,\ell=1}^{3} \epsilon_{jk\ell} < a_j | 1 > (\bar{a}_k - a_k) b_\ell , \tag{8.41c}$$

where

$$\bar{a}_k = 2 < a_k | 1 > 1 - a_k ,$$

as usual. Here $\mu \in F$ in Eq. (8.41b) is an arbitrary, but fixed, constant. Then $[x_1, x_2, x_3]$ defined above is clearly totally anti-symmetric in x_1, x_2, and x_3. Moreover, it defines an alternative triple system, with $B_z(x, y)$ being given again by Eq. (8.25) if we identify

$$(x_1 | x_2) = < a_1 | a_2 > - \mu < b_1 | b_2 > , \tag{8.42a}$$

$$e = 1 \oplus 0 , \tag{8.42b}$$

$$\psi(x) = (e | x) . \tag{8.42c}$$

Then the product $x_1 \cdot x_2$ with $\lambda = 1$ and $\phi(x) = \psi(x)$ is calculated to be

$$x_1 \cdot x_2 = a' \oplus b' , \tag{8.43a}$$

$$a' = a_1 a_2 + \mu b_2 \bar{b}_1 , \tag{8.43b}$$

$$b' = \bar{a}_1 b_2 + a_2 b_1 . \tag{8.43c}$$

If we now choose Dim $B = 2$, this process will give the four-dimensional quaternion algebra for V, while another choice of Dim $B = 4$ (i.e. B being the quaternion algebra) will lead to an octonion algebra.

(c) Octonion as 3 × 3 traceless matrices

Let X, Y, Z be 3×3 traceless matrices with the associative matrix product XY and set

$$[X, Y, Z] = \frac{1}{2} \{X[Y, Z] + Y[Z, X] + Z[X, Y]\} - \frac{1}{2} \text{Tr} (X[Y, Z])E , \tag{8.44}$$

where E is the 3×3 unit matrix. Since we have Tr $([X, Y, Z]) = 0$, the eight-dimensional vector space V of all 3×3 traceless matrices admits a totally anti-symmetric triple-linear product $[X, Y, Z]$. Moreover, recall the fact that X satisfies a cubic equation as in Section 4.2. Utilizing these, we can prove the validity of

$$[X,[X,Y,Z],Z] = B_Z(X,Y)X + B_X(Z,Y)Z - B_Z(X,X)Y \ , \qquad (8.45a)$$

with

$$B_Z(X,Y) = \frac{1}{4} \operatorname{Tr} (XZ) \operatorname{Tr} (YZ) - \frac{1}{2} \operatorname{Tr} (XZYZ) \ . \qquad (8.45b)$$

Therefore, the triple system is alternative. However, since $B_Z(X,Y)$ given above has a form like neither Eq. (8.24) nor Eq. (8.25), we must be careful of the choice of $\phi(x)$ and e so as to ensure the non-degeneracy of $<X|Y>$ constructed in Theorem 12. Let Λ be any 3×3 traceless matrix such that $\operatorname{Tr} \Lambda^3 \neq 0$, and set

$$\phi(X) = \frac{\operatorname{Tr} (X\Lambda^2)}{\operatorname{Tr} \Lambda^3} \ , \qquad (8.46)$$

which satisfies $\phi(e) = 1$ for the choice of $e = \Lambda$. Now, setting

$$X \cdot Y = \phi(X)Y + \phi(Y)X - <X|Y> \Lambda + \lambda\{[X,Y,\Lambda] - \phi([X,Y,\Lambda])\Lambda\} \ , \qquad (8.47)$$

this defines the octonion algebra, since

$$<X|Y> = \phi(X)\phi(Y) + \frac{1}{4} \lambda^2 \{\operatorname{Tr} (X\Lambda) \operatorname{Tr} (Y\Lambda) - 2 \operatorname{Tr} (X\Lambda Y\Lambda)\} \qquad (8.48)$$

for $\lambda \neq 0$ can be shown to be non-degenerate. Perhaps the simplest choice for the traceless matrix Λ satisfying $\operatorname{Tr} \Lambda^3 \neq 0$ would be

$$\Lambda = \begin{pmatrix} 1 & 0 & 0 \\ 0 & 1 & 0 \\ 0 & 0 & -2 \end{pmatrix} = \sqrt{3} \, \lambda_8 \ .$$

Octonion algebra can be rewritten in terms of the Gell–Mann symbols $f_{jk\ell}$ and $d_{jk\ell}$ as for pseudo-octonion algebra. We can then derive the multiplication table of Chapter 1 again. For details, see reference 63.

Although we can construct other triple systems based upon a_μ and $f_{\mu\nu}$ of Chapter 1 as well as the spin $J = 3$ system of Chapter 6, we will not go into the details here.

Remark 8.5. There is another way of expressing the octonion as traceless 3×3 matrices as follows. Let $X * Y$ be the product of pseudo-octonion algebra defined by Eq. (4.9). For any traceless matrix B satisfying

$< B|B > \neq 0$, we have

$$(R_B)^{-1}X = \frac{1}{< B|B >} \, B * X \, ,$$

$$(L_B)^{-1}X = \frac{1}{< B|B >} \, X * B \, ,$$

in view of relations (4.13) and (4.14). The new product

$$X \circ Y = (R_A^{-1}X) * (L_A^{-1}Y) = \left(\frac{1}{< B|B >}\right)^2 (B * X) * (Y * B)$$

defines an octonion algebra with the unit element $e = B * B$ because of the result explained in Remark 3.5 of Chapter 3. Expressing $X * Y$ as the matrix product by Eq. (4.9) gives the desired answer.

8.4 Octonionic triple system

Let L and $V = [1]$ be a Lie algebra and an irreducible L-module respectively. Suppose that we have

$$\text{Dim Hom} \, ([2] \rightarrow F) = 1 \, , \tag{8.49a}$$

$$\text{Dim Hom} \, ([1^3] \rightarrow [1]) = 1 \, , \tag{8.49b}$$

$$\text{Dim Hom} \, ([1^4] \rightarrow F) = 1 \, , \tag{8.49c}$$

$$\text{Dim Hom} \, ([1^3] \otimes [1^3] \rightarrow F) \leq 2 \, . \tag{8.49d}$$

As in Chapter 6, Eq. (8.49a) implies the existence of the unique (apart from overall normalization) symmetric bi-linear non-degenerate form $< x|y >$. Next, there exists a unique, totally anti-symmetric triple linear product $[x, y, z]$ because of Eq. (8.49b). Just as we derived Eqs. (6.28) from relation (6.26) we can now infer the validity of the following from Eq. (8.49c):

$$< w|[x, y, z] > = \text{ totally anti-symmetric in } w, \ x, \ y, \text{ and } z \, . \tag{8.50}$$

Finally, by the same reasoning as in Chapter 6 where we obtained the relation Eq. (6.33), Eq. (8.49d) leads to

$$< [x, y, z]|[u, v, w] > = \alpha \sum_P (-1)^P < x|u >< y|v >< z|w >$$

$$+ \frac{1}{4} \beta \sum_P \sum_{P'} (-1)^P (-1)^{P'} < x|u >$$

$$\cdot < y|[z, v, w] > \tag{8.51}$$

for some constants α and β, since the left side of Eq. (8.51), as well as the coefficients of α and β, are three elements of Hom $([1]^3 \otimes [1]^3 \rightarrow F)$. Here, P and P' refer to 3! permutations of x, y, and z and of u, v, and w respectively.

Next, we rewrite the left side of Eq. (8.51) as $- < w|[u, v, [x, y, z]] >$ by Eq. (8.50) and note the non-degeneracy of $< w|s >$. Then Eq. (8.51) is equivalent to a triple product relation generalizing Eq. (8.9),

$$
\begin{aligned}
[u,v, &[x, y, z]] \\
= &\{\alpha(< y|v >< z|u > \; - \; < y|u >< z|v >) - \beta < u|[v, y, z] >\}x \\
&+ \{\alpha(< z|v >< x|u > \; - \; < z|u >< x|v >) - \beta < u|[v, z, x] >\}y \\
&+ \{\alpha(< x|v >< y|u > \; - \; < x|u >< y|v >) - \beta < u|[v, x, y] >\}z \\
&- \beta\{< x|v > [u, y, z] + \; < y|v > [u, z, x] + < z|v > [u, x, y] \\
&+ < x|u > [v, z, y] + \; < y|u > [v, x, z] + \; < z|u > [v, y, x]\} \; .(8.52)
\end{aligned}
$$

Setting $u = x$ and $v = z$ in Eq. (8.52), we then find

$$[x, [x, y, z], z] = B_z(y, x)x + B_x(y, z)z - B_z(x, x)y \, , \qquad (8.53a)$$

$$B_z(x, y) = \alpha(< x|y >< z|z > \; - \; < z|y >< x|z >) \, , \qquad (8.53b)$$

which reproduce Eqs. (8.11) and (8.12). Note, especially, that $B_z(x, y)$ possesses the form specified by Eq. (8.24). Therefore, if $\alpha \neq 0$, then we can construct the quaternion algebra for the case of Dim $V = 4$, and the octonion algebra if Dim $V = 8$. Without detailed calculations, we simply mention that Eqs. (8.49) can be satisfied for two cases of

(i) $L =$ so(4), Dim $V = 4$, with V being the vector representation,

(ii) $L =$ so(7), Dim $V = 8$, with V being the spinor representation,

as we have noted elsewhere.[66]

Moreover, Eqs. (8.52) with Eq. (8.50) can be shown,[66] to be self-consistent for only the two cases:

(i) Dim $V = 4$, $\alpha \neq 0$, $\beta = 0$, $\qquad\qquad\qquad$ (8.54a)

(ii) Dim $V = 8$, $\alpha = \beta^2 \neq 0$, $\qquad\qquad\qquad$ (8.54b)

provided that $[u, v, [x, y, z]]$ is not identically zero. In order to prove this statement, let Dim $V = N$ with a basis $e_1, e_2, \ldots, e_N$ of the vector space V. We now set

$$g_{jk} = g_{kj} = < e_j|e_k > \, , \qquad (8.55a)$$

$$e^j = g^{jk}e_k \, , \qquad (8.55b)$$

where the automatic summation convention on the repeated indices is understood here, and hereafter. We obviously have the identity

$$e_j < e^j|x> = <x|e_j> e^j = x .$$ (8.56)

After these preparations, consider the expression

$$J = [u, [a, b, c], [x, y, z]]$$ (8.57)

for u, a, b, c, x, y, $z \in V$. Setting $v = [a, b, c]$, we can calculate

$$J = [u, v, [x, y, z]]$$

by Eq. (8.52). But a term such as

$$[v, x, y] = [x, y, v] = [x, y, [a, b, c]]$$

can again be written in a simpler form using Eq. (8.52) for $u \rightarrow x \rightarrow a$, $v \rightarrow y \rightarrow b$, and $z \rightarrow c$. In this way, we can reduce J to a complicated, but simpler, expression, involving only single triple products. However, Eq. (8.57) requires that J will change its sign for the interchanges of $a \leftrightarrow x$, $b \leftrightarrow y$, and $c \leftrightarrow z$. These two facts lead to a rather complicated identity, whose explicit form will not be given here. We consider several special cases of the identity by setting

$$\text{(a)} \quad c = e_j \text{ and } z = e^j$$ (8.58a)

or

$$\text{(b)} \quad u = e_j \text{ and } c = e^j$$ (8.58b)

and summing over the repeated index j. Using Eq. (8.56), we can then prove the validity of Eqs. (8.54). For details, see reference 66.

Hereafter, we will restrict ourselves to the case where $N = 8$, unless stated otherwise. In view of Eq. (8.54b), we may assume that

$$\alpha = -\beta = 1 ,$$ (8.59)

without loss of generality, when we normalize $< x|y >$ and/or $[x, y, z]$ suitably. For the choice of $\phi(x) = < e|x >$, we now note that

$$\phi([e, x, y]) = 0$$ (8.60)

since we have, by Eq. (8.50),

$$\phi([e, x, y]) = < e|[e, x, y] > = 0 .$$

Therefore, with $\lambda = 1$,

$$x \cdot y = [x, y, e] + < x|e > y + < y|e > x - < x|y > e$$ (8.61)

will define an octonion algebra by the result of Theorem 12, together with Remark 8.3. Actually, the validity of the composition law

$$< x \cdot y | x \cdot y > \; = \; < x | x > < y | y > \tag{8.62}$$

can be verified more directly from Eqs. (8.51) and (8.50).

Another interesting fact of the octonionic triple system is that we can express $[x, y, z]$ conversely in terms of the octonionic product $x \cdot y$ as follows. Setting $z = v = e$ in Eq. (8.52) and then replacing u by z, we obtain

$$[x, y, z] = (x \cdot y) \cdot z - 2 < y | e > x \cdot z$$
$$+ < y | z > x + < x | y > z - < x | z > y . \tag{8.63}$$

Anti-symmetrizing both sides of Eq. (8.63), we can rewrite it in the symmetrical form

$$[x, y, z] = \frac{1}{2} \{ (x, y, z) + < x | e > [y, z]$$
$$+ < y | e > [z, x] + < z | e > [x, y] - < z | [x, y] > e \} , \tag{8.64}$$

where (x, y, z) and $[x, y]$ are the associator and commutator, respectively, given by

$$(x, y, z) = (x \cdot y) \cdot z - x \cdot (y \cdot z)$$
$$[x, y] = x \cdot y - y \cdot x .$$

Here we have utilized the identity Eq. (2.19). We also note that expression (8.64) differs in form from Eq. (8.28a) for both the Zorn and Cayley–Dickson cases. Because of this reciprocity between the triple product and the octonion algebra, we call the present triple system the octonionic triple system. However, we note that the derivation algebra of the octonionic triple product is at least so(7) (actually it is so(8)), which is larger than G_2 of the octonion algebra.

An explicit construction of the octonionic triple system can be performed as follows. If the field F is complex, we may consider the unique eight-dimensional realization of the complex Clifford algebra in seven-dimensional vector space. If we are considering the real field F, we choose the eight-dimensional normal representation of the real Clifford algebra $C(4, 3)$ or $C(0, 7)$, as in Theorem 7 of Chapter 5. At any rate, the underlying vector space is the eight-dimensional spinor representation of so(7) or its non-compact form so(4,3). There exists[9,25] a (complex or real) charge conjugation matrix C satisfying

$$C \gamma_\mu C^{-1} = -\gamma_\mu^{\mathrm{T}} , \quad C^{\mathrm{T}} = C . \tag{8.65}$$

Let ξ_1, ξ_2, and ξ_3 be three generic eight-dimensional spinors and set

$$< \xi_1|\xi_2 > = \xi_1^{\mathrm{T}} C \xi_2 , \tag{8.66}$$

$$
[\xi_1, \xi_2, \xi_3] = \frac{1}{3} \sum_{\mu=1}^{7} \{ \gamma_\mu \xi_1 (\xi_2^{\mathrm{T}} C \gamma^\mu \xi_3)
$$
$$
+ \gamma_\mu \xi_2 (\xi_3^{\mathrm{T}} C \gamma^\mu \xi_1) + \gamma_\mu \xi_3 (\xi_1^{\mathrm{T}} C \gamma^\mu \xi_2) \} , \tag{8.67}
$$

where ξ^{T} denotes the transpose of the spinor ξ, regarded as an eight-dimensional vector in V. Clearly, then, $< \xi_1|\xi_2 > = < \xi_2|\xi_1 >$ is a symmetric bi-linear non-degenerate form in V, and $[\xi_1, \xi_2, \xi_3]$ is a totally anti-symmetric triple linear product in V. Using the Fierz identity[9] based upon the orthogonality relation, Eq. (5.14), we can then prove[25] that our triple product $[\xi_1, \xi_2, \xi_3]$ satisfies the desired relation Eq. (8.52), with $\alpha = -\beta = 1$. However, the details of the calculation are not given here.

Finally, it may be worth while remarking on the structure constants of our triple product. Let $e_1, e_2, \ldots, e_N$ for either $N = 4$ or 8 be a basis of the quaternionic or octonionic triple system, respectively, and introduce g_{jk} and e^j by Eqs. (8.55). We can then express

$$[e_i, e_j, e_k] = \sum_{\ell=1}^{N} C_{ijk\ell} e^\ell \tag{8.68}$$

for some constant $C_{ijk\ell}$, which can be computed conversely by

$$C_{ijk\ell} = < e_\ell|[e_i, e_j, e_k] > . \tag{8.69}$$

Then $C_{ijk\ell}$ is totally anti-symmetric in the four-indices i, j, k, and ℓ because of Eq. (8.50). Moreover, Eq. (8.51) or Eq. (8.52) gives

$$
\sum_{p,q=1}^{N} g^{pq} C_{abcp} C_{ijkq} = \alpha \sum_P (-1)^P g_{ai} g_{bj} g_{ck}
$$
$$
+ \frac{1}{4} \beta \sum_P \sum_{P'} (-1)^P (-1)^{P'} g_{ai} C_{bcjk} , \tag{8.70}
$$

which reproduces the relation found by de Wit and Nicolai,[67] and by Gürsey and Tze,[68] who constructed $C_{ijk\ell}$ from the octonion algebra by using formula (8.64).

Remark 8.6. Relation (8.70) has been utilized by de Wit and Nicolai,[67] as well as by Gürsey and Tze,[68] in their discussions of an 11-dimensional Kaluza–Klein theory.

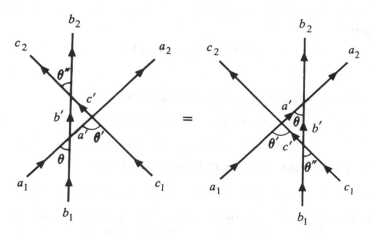

Fig. 8.1 Yang–Baxter equation.

8.5 Application to Yang–Baxter equation

(a) Yang–Baxter equation and new solution method

The Yang–Baxter (Y–B) equation[69]

$$R_{a_1b_1}^{b'a'}(\theta)R_{a'c_1}^{c'a_2}(\theta')R_{b'c'}^{c_2b_2}(\theta'')$$
$$= R_{b_1c_1}^{c'b'}(\theta'')R_{a_1c'}^{c_2a'}(\theta)R_{a'b'}^{b_2a_2}(\theta) , \qquad (8.71a)$$

where

$$\theta' = \theta + \theta'' \qquad (8.71b)$$

appears in many subjects[69-72] from statistical mechanics, exactly solvable two-dimensional field theory, the quantum group, Hopf algebra, and the braid group. Here, and hereafter, all repeated indices are understood to be automatically summed over the N values $1, 2, \ldots, N$ for some integer N. A simple way of visualizing Eqs. (8.71) is to regard $R_{ab}^{a'b'}(\theta)$ as a scattering matrix for the reaction

$$a + b \rightarrow a' + b' ,$$

with rapidity parameter θ in one-dimensional space, where a, b, a', and b' refer to indices such as spins for N possible states for each particle. Then Eq. (8.71a) can be depicted graphically as in Fig. 8.1, where the vertical direction indicates the flow of the time, and where the horizontal line refers to one-dimensional space.

It is often more convenient to consider an N-dimensional vector space V (normally real or complex) with basis vectors $e_1, e_2, \ldots, e_N$. We ordinarily introduce the general linear transformations $R_{12}(\theta)$, $R_{13}(\theta)$, and $R_{23}(\theta)$:

$$V \otimes V \otimes V \longrightarrow V \otimes V \otimes V$$

by

$$R_{12}(\theta)\{e_a \otimes e_b \otimes e_c\} = R_{ab}^{b'a'}(\theta)e_{a'} \otimes e_{b'} \otimes e_c ,$$
$$R_{23}(\theta)\{e_a \otimes e_b \otimes e_c\} = R_{bc}^{c'b'}(\theta)e_a \otimes e_{b'} \otimes e_{c'} , \quad (8.72)$$
$$R_{13}(\theta)\{e_a \otimes e_b \otimes e_c\} = R_{ac}^{c'a'}(\theta)e_{a'} \otimes e_b \otimes e_{c'} .$$

Then the Y–B equation, Eq. (8.71a), is rewritten as

$$R_{12}(\theta)R_{13}(\theta')R_{23}(\theta'') = R_{23}(\theta'')R_{13}(\theta')R_{12}(\theta) , \quad (8.73)$$

which is convenient for discussion of the underlying structure of the theory, such as the Hopf algebra. However, it is not easy in general to find an explicit solution of Eq. (8.73), except for some simple systems.

We will now present a new way of solving the Y–B equation in terms of some triple product systems. We first assume that the vector space V possesses a non-degenerate bi-linear form $< x|y >$ which is not yet assumed to be symmetric. We now set

$$g_{ab} = < e_a|e_b > \quad (8.74)$$

and write its inverse as g^{ab}, so that

$$g^{ac}g_{cb} = g_{bc}g^{ca} = \delta_b^a , \quad < e^a|e_b > = \delta_b^a . \quad (8.75)$$

Moreover, we introduce e^a by

$$e^a = g^{ab}e_b \quad \text{or} \quad e_a = g_{ab}e^b , \quad (8.76)$$

which satisfies Eq. (8.56), that is,

$$e_a < e^a|x > = < x|e_a > e^a = x . \quad (8.77)$$

Next, we assume for simplicity that $R_{ab}^{cd}(\theta)$ satisfies the symmetric condition

$$R_{ab}^{cd}(\theta) = R_{ba}^{dc}(\theta) , \quad (8.78)$$

and define a θ-dependent triple product

$$[x, y, z]_\theta : V \otimes V \otimes V \longrightarrow V \quad (8.79)$$

by

$$[e^d, e_a, e_b]_\theta = e_c \, R_{ab}^{cd}(\theta) , \quad (8.80)$$

or, equivalently, by

$$< e^c | \left[e^d, e_a, e_b \right]_\theta > = R_{ab}^{cd}(\theta) . \tag{8.81}$$

Then Eq. (8.78) is rewritten as

$$< u | [z, x, y]_\theta > = < z | [u, y, x]_\theta > \tag{8.82}$$

while Eq. (8.71a) becomes a triple product equation[66]

$$
\begin{aligned}
[v, [u, e_j, z]_{\theta'} , \ [e^j, x, y]_\theta]_{\theta''} \\
= [u, [v, e_j, x]_{\theta'} , \ [e^j, z, y]_{\theta''}]_\theta
\end{aligned}
\tag{8.83}
$$

in a basis-independent notation. Indeed, if we set $x = e_{a_1}$, $y = e_{b_1}$, $z = e_{c_1}$, $u = e^{a_2}$, and $v = e^{c_2}$ in Eq. (8.83), it will reproduce the Y–B equation (8.71a). Therefore, our task is reduced to finding a triple product $[x, y, z]_\theta$ which satisfies both Eqs. (8.82) and (8.83).

To illustrate our method, consider the following case. We assume that $< x | y >$ is either symmetric or anti-symmetric, that is,

$$< x | y > = \epsilon < y | x > , \tag{8.84a}$$

$$\epsilon = +1 \text{ or } -1 , \tag{8.84b}$$

and seek a solution in the form of

$$[z, x, y]_\theta = A(\theta) < x | y > z + B(\theta) < z | x > y + C(\theta) < y | z > x \tag{8.85}$$

for scalar functions $A(\theta)$, $B(\theta)$, and $C(\theta)$ to be determined. In view of Eq. (8.84), condition (8.82) can easily be seen to be automatically satisfied by Eq. (8.85). Inserting Eq. (8.85) into (8.83), we can verify that we need to solve only the following three equations:

$$B''C'C + C''C'B - C''B'C = 0 , \tag{8.86a}$$
$$B''A'C - A''B'C - A''A'B = 0 , \tag{8.86b}$$

and

$$
\begin{aligned}
\epsilon N A''C'A + (A''A' + A''B' + B''C' + C''C')A \\
+ (A''C' - C''A')C + A''C'B = 0 ,
\end{aligned}
\tag{8.86c}
$$

where for simplicity we have set

$$A = A(\theta) , \quad A' = A(\theta') , \quad A'' = A(\theta'') , \tag{8.87}$$

etc. In view of the condition $\theta' = \theta + \theta''$, as in Eq. (8.71b), these are coupled function equations whose complete solutions are given below:

(i) $A(\theta) = C(\theta) = 0$, $B(\theta) =$ arbitrary ,   (8.88a)

(ii) $A(\theta) = B(\theta) = 0$, $C(\theta) =$ arbitrary ,   (8.88b)

(iii) $B(\theta) = 0$, $\frac{C(\theta)}{A(\theta)} = \frac{1}{2} \left\{ \pm\sqrt{N^2 - 4} \, \frac{e^{k\theta}+1}{e^{k\theta}-1} - \epsilon N \right\}$,   (8.88c)

(iv) $B(\theta) = 0$, $\frac{C(\theta)}{A(\theta)} = \frac{1}{2} \left\{ \pm\sqrt{N^2 - 4} - \epsilon N \right\}$,   (8.88d)

(v) $B(\theta) = 0$, $N = 2$, $\frac{C(\theta)}{A(\theta)} = \frac{1}{k\theta} - \epsilon$,   (8.88e)

(vi) $\frac{B(\theta)}{C(\theta)} = k\theta$, $\frac{C(\theta)}{A(\theta)} = -1 - \frac{\epsilon N - 2}{2k\theta}$,   (8.88f)

where k is an arbitrary constant. Solution (8.88c) is known to be of a trigonometric type, while solutions (8.88e) and (8.88f) are said to be of a rational type. Also, Eqs. (8.88a), (8.88b), and (8.88d) correspond to constant solutions.

We will give a derivation of solution (8.88f) only, as follows. Setting

$$f(\theta) = \frac{B(\theta)}{C(\theta)} ,$$

then Eq. (8.86a) can be rewritten as

$$f(\theta) + f(\theta'') = f(\theta') = f(\theta + \theta'') ,$$

whose solution must be of the form of

$$\frac{B(\theta)}{C(\theta)} = f(\theta) = k\theta$$

for a constant k. Consider, next, Eq. (8.86b), which can be rewritten as

$$\frac{B''}{C''}\frac{A'}{C'} - \frac{A''}{C''}\frac{B'}{C'} - \frac{A''}{C''}\frac{A'}{C'}\frac{B}{C} = 0 ,$$

which can be solved for $k \neq 0$ to give

$$\frac{A(\theta)}{C(\theta)} = \frac{k\theta}{\alpha - k\theta}$$

for a constant α. Inserting these results into Eq. (8.86c), α is determined to be

$$\alpha = -\frac{1}{2} (\epsilon N - 2) ,$$

which gives the desired solution (8.88f).

Many of the solutions given in Eqs. (8.88) have been previously found by various authors for the case of $\epsilon = +1$.[69,73] We can also find more complicated solutions of the Y–B equations. As an example, let $[x, y, z]$ be the totally anti-symmetric octonionic or quaternionic triple product discussed in the preceding section, 8.4, with the symmetric bi-linear product $< y|x > = < x|y >$ corresponding to $\epsilon = +1$, thereby restricting ourselves to the two cases $N = 8$ or 4.

We seek a solution in the form

$$[z, x, y]_\theta = P(\theta)[x, y, z] + A(\theta) < x|y > z$$
$$+ B(\theta) < z|x > y + C(\theta) < y|z > x \qquad (8.89)$$

for another function $P(\theta) \neq 0$. Note that constraint (8.82) is also automatically satisfied by Eq. (8.89) when we use Eq. (8.50). Inserting Eq. (8.89) into both sides of Eq. (8.83), and using Eqs. (8.52) and (8.56), we now find that P, A, B, and C must satisfy seven sets of complicated cubic equations instead of Eqs. (8.86). However, the solutions have been found and are given as follows. For simplicity, we choose $\alpha = -\beta = 1$ for $N = 8$, while we set $\alpha = 1$ and $\beta = 0$ for $N = 4$ by the result of the preceding section. Then we obtain the rational-type solutions:

(a) N = 8

$$\frac{A(\theta)}{P(\theta)} = \frac{18 - 3b\theta}{10 - b\theta} , \qquad (8.90a)$$

$$\frac{B(\theta)}{P(\theta)} = b\theta - 5 , \qquad (8.90b)$$

$$\frac{C(\theta)}{P(\theta)} = \frac{12 - 3b\theta}{b\theta} , \qquad (8.90c)$$

where b is an arbitrary constant. If we choose $b = -3$ with the normalization $C(\theta) = \frac{1}{\theta}$, this reproduces de Vega and H. Nicolai's result.[74]

(b) N = 4

$$\frac{A(\theta)}{P(\theta)} = -a , \qquad (8.91a)$$

$$\frac{B(\theta)}{P(\theta)} = a + b\theta , \qquad (8.91b)$$

$$\frac{C(\theta)}{P(\theta)} = a + \frac{a^2 - 1}{b\theta} , \qquad (8.91c)$$

where a and b are arbitrary constants.

We will, however, omit detailed calculations. Also, we can generalize the method to find solutions for more complicated triple systems. See reference 66 for more details.

8.6 Nambu's dynamical equation and Lie triple system

Among many proposed methods for generalizing or modifying the present framework of quantum mechanics, Nambu's proposal to modify the Heisenberg equation of motion, Eq. (7.5), into a triple product equation is of interest for the theory of the triple product. Nambu suggested[75] replacing Eq. (7.5) by

$$\frac{dx}{dt} = \{h_1, h_2, x\} ,$$ (8.92)

where $\{x, y, z\}$ is a triple-linear product, and we use two Hamiltonian operators h_1, and h_2, instead of the customary one Hamiltonian as given in Eq. (7.5). If $\{x, y, z\}$ is totally anti-symmetric in x, y, and z, then Eq. (8.92) implies

$$\frac{d}{dt} h_1 = \frac{d}{dt} h_2 = 0 ,$$

so that both h_1 and h_2 are conserved. However, here we will not necessarily assume total anti-symmetricity for the triple product unless so stated. Hence, h_1 and h_2 may not necessarily be conserved quantities of the system. Now, recall the argument presented in Chapter 7. Suppose that x, y, and z are physical observables in some sense. It would then be natural to assume the same for $\{x, y, z\}$. Moreover, it is reasonable to assume the validity of the Leibnitz rule

$$\frac{d}{dt}\{x, y, z\} = \left\{\frac{dx}{dt}, y, z\right\} + \left\{x, \frac{dy}{dt}, z\right\} + \left\{x, y, \frac{dz}{dt}\right\} .$$ (8.93)

Replacing x by $\{x, y, z\}$ in Eq. (8.92) and using Eq. (8.93), we obtain the consistency condition

$$\{h_1, h_2, \{x, y, z\}\} = \{\{h_1, h_2, x\}, y, z\} \\ + \{x, \{h_1, h_2, y\}, z\} + \{x, y, \{h_1, h_2, z\}\}$$ (8.94)

for any three observables x, y, and z. It would be natural, as in Chapter 7, to assume the validity of the stronger assumption that any five observables u, v, x, y, and z would satisfy

$$\{u, v, \{x, y, z\}\} = \{\{u, v, x\}, y, z\}$$
$$+ \{x, \{u, v, y\}, z\} + \{x, y, \{u, v, z\}\} \ . \qquad (8.95)$$

We note, in particular, that if H_1, H_2, H_3 are conserved quantities, that is, $\frac{d}{dt} H_1 = \frac{d}{dt} H_2 = \frac{d}{dt} H_3 = 0$, then so will be $H = \{H_1, H_2, H_3\}$, that is, $\frac{d}{dt} H = 0$ because of Eq. (8.93). Now define a linear mapping $D_{u,v}$, $V \to V$ by

$$D_{u,v} \ x = \{u, v, x\} \ . \qquad (8.96)$$

Equation (8.95) is rewritten as

$$D_{u,v}\{x, y, z\} = \{D_{u,v}x, y, z\} + \{x, D_{u,v}y, z\} + \{x, y, D_{u,v}z\} \ . \qquad (8.97)$$

On comparison with Eq. (8.5), this implies that $D_{u,v}$ is a derivation of the triple product $\{x, y, z\}$. Moreover, we can rewrite Eq. (8.95) as

$$\left(D_{u,v}D_{x,y} - D_{x,y}D_{u,v}\right) z = D_{\{u,v,x\},y}z + D_{x,\{u,v,y\}}z \ ,$$

or, equivalently,

$$\left[D_{u,v}, D_{x,y}\right] = D_{\{u,v,x\},y} + D_{x,\{u,v,y\}} \ , \qquad (8.98)$$

so that a set consisting of all $D_{x,y}$s forms a derivation Lie algebra of a special form.

The next question is whether there exist triple product systems satisfying Eq. (8.95) or, equivalently, Eq. (8.98). The answer is affirmative. Many interesting triple systems do satisfy Eq. (8.95), as we will show in Remark 8.8 below.

Returning to our discussion of the Nambu equation, Eq. (8.92), Nambu further assumed the existence of the standard bi-linear product xy in addition to the triple product. The consistency condition (7.6), that is, $\frac{d}{dt} (xy) = \frac{dx}{dt} y + x \frac{d}{dt} y$ also requires the validity of

$$\{h_1, h_2, xy\} = \{h_1, h_2, x\} y + x \{h_1, h_2, y\} \ .$$

Again, it is tempting to assume the validity of the stronger assumption that any four observables u, v, x, and y will satisfy

$$\{u, v, xy\} = \{u, v, x\}y + x\{u, v, y\} \ , \qquad (8.99)$$

which is rewritten as

$$D_{u,v}(xy) = \left(D_{u,v}x\right) y + x \left(D_{u,v}y\right) \ . \qquad (8.100)$$

In other words, $D_{u,v}$ is also a derivation of the bi-linear product xy, as well as that of the triple product $\{x, y, z\}$.

Conversely, let A be an algebra with a bi-linear product xy. Assume that for any two $u, v \in A$, we can find a derivation $D_{u,v}$ of A such that it is bi-linear in u and v. Then we can introduce a triple product $\{u, v, x\}$ by Eq. (8.96). It generally satisfies Eq. (8.95), as we see from the following example. Consider, for example, the alternative algebra. We have seen that $D_{x,y}$, given by Eq. (3.68), is a derivation of alternative algebra so that

$$\{x, y, z\} = D_{x,y}z = \left\{ [L_x, L_y] + [R_x, R_y] + [L_x, R_y] \right\}z$$
$$= x(yz + xy) - (zx + xz)y + (zy)x - y(xz) . \quad (8.101)$$

Since $D_{u,v}$ is a derivation of A, it also implies the validity of Eq. (8.95) when $D_{u,v}$ is applied once more to both sides of Eq. (8.101). However, $\{x, y, z\}$, introduced by Eq. (8.101), is not totally anti-symmetric in x, y, z.

A more illuminating case will also be obtained as follows. Let A be an associative algebra with a bi-linear associative product xy. Setting $[x, y] = xy - yx$ as usual, we can now introduce a triple product by

$$\{x, y, z\} = [[x, y], z] . \quad (8.102)$$

The Nambu equation will then reduce to the Heisenberg equation of motion, Eq. (7.5), if we identify

$$H = -i\,[h_1, h_2] . \quad (8.103)$$

The triple product defined by Eq. (8.102) satisfies the following relations:

$$\{x, y, z\} = -\{y, x, z\} , \quad (8.104a)$$
$$\{x, y, z\} + \{y, z, x\} + \{z, x, y\} = 0 , \quad (8.104b)$$

in addition to the derivation property Eq. (8.95). More generally, any triple system satisfying Eqs. (8.95) and (8.104) is called a Lie-triple system.[76,77] Let V be a Lie-triple system. Then the left-multiplication operator $D_{x,y} \in g\ell(V)$ introduced by Eq. (8.96) satisfies Eq. (8.98), that is,

$$[D_{u,v}, D_{x,y}] = D_{\{u,v,x\},y} + D_{x,\{u,v,y\}} . \quad (8.98')$$

Next, let M be a sub-vector space of $g\ell(V)$, which is spanned by all $D_{u,v} \in g\ell(V)$. Following Lister[76] and Yamaguchi,[77] we define a new anti-symmetric product $[X, Y]$ in $V_0 = M \oplus V$ by

$$[x, y] = D_{x,y} , \quad (8.105a)$$
$$[D_{x,y}, z] = \{x, y, z\} , \quad (8.105b)$$

for any x, y, $z \in V$. Now, we clearly have

$$\{x, y, z\} = [[x, y], z] , \tag{8.106}$$

by Eqs. (8.105). Moreover, we can readily verify that Eqs. (8.98) and (8.105) define a Lie algebra. For example, we compute

$$[[x, y], z] + [[y, z], x] + [[z, x], y] = \{x, y, z\} + \{y, z, x\} + \{z, x, y\} = 0$$

by Eq. (8.104b) and Eq. (8.106). Applying Eq. (8.106) to the Nambu equation, we recover the Heisenberg equation of motion, Eq. (7.5), again, with H being given by Eq. (8.103). Moreover, we note that $[M, M] \subset M$, $[M, V] \subset V$, but that $[V, V] \subset M$ from Eqs. (8.98'), and (8.105), which is familiar in the theory of symmetric homogeneous space.[78]

As another application of a Lie-triple system, let $< x|y >$ be a symmetric bi-linear form and consider

$$\{x, y, z\} = \lambda \{< y|z > x - < z|x > y\} \tag{8.107}$$

for a constant λ. It is then easy to verify that this defines a Lie-triple system. Suppose now that the underlying vector space V consists of fermionic annihilation operators a_j and creation operators a_j^+ ($j = 1, 2, \ldots, N$), and assign

$$< a_j|a_k^+ > = < a_k^+|a_j > = \delta_{jk} , \tag{8.108a}$$
$$< a_j|a_k > = < a_j^+|a_k^+ > = 0 . \tag{8.108b}$$

Then, with the choice of $\lambda = 2$, we have relations such as

$$\{a_j, a_k^+, a_\ell\} = -\{a_k^+, a_j, a_\ell\} = 2\, \delta_{k\ell} a_j , \tag{8.109a}$$
$$\{a_j, a_k, a_\ell^+\} = 2\, (\delta_{k\ell} a_j - \delta_{j\ell} a_k) , \tag{8.109b}$$

etc. Together with Eq. (8.106), we can rewrite Eq. (8.109a), for example, as

$$[[a_j, a_k^+], a_\ell] = 2\, \delta_{k\ell} aj ,$$

etc., and we recognize this as a relation of the para-fermionic system.[79] For a para-boson case, we need the notion of a Lie-super triple system,[80] although we will not go into the details here.

Remark 8.7. As we noted at the beginning of this section, the most interesting case for Nambu dynamics is when the triple product $\{x, y, z\}$ is totally anti-symmetric in x, y, and z, instead of it being a Lie-triple system, since both h_1 and h_2 are then conserved. Unfortunately, it is not easy to find such a system, other than the four-dimensional quaternionic triple system, as we will see below. Also, an example of infinite-dimensional Nambu dynamics has been given by Bialynicki–Birula and Morrison in reference 81.

Remark 8.8. Let $[x, y, z]$ be the totally anti-symmetric quaternionic or octonionic triple system discussed in Section 8.4. Then we can construct a new triple system $\{x, y, z\}$ by

$$\{x, y, z\} = [x, y, z] + \lambda < y|z > x - \lambda < z|x > y, \qquad (8.110)$$

which satisfies the derivation property Eq. (8.95), as well as

$$\{x, y, z\} + \{y, x, z\} = 0,$$
$$\{x, y, z\} + \{x, z, y\} = 2\lambda < y|z > x - \lambda < x|y > z - \lambda < z|x > y. \qquad (8.111)$$

Here, λ is an arbitrary constant for the quaternionic case of $N = 4$, but we must choose it to be

$$\lambda = -3\,\beta \qquad (8.112)$$

for the octonionic case of $N = 8$. For the quaternionic case in particular, we may choose $\lambda = 0$ with $\{x, y, z\} = [x, y, z]$. We say that any triple system satisfying Eqs. (8.95) and (8.111) is an orthogonal triple system, which is a supersymmetric analog of the so-called symplectic triple system, studied by some mathematicians.[82] Both orthogonal and symplectic triple systems have also been used to find solutions of the Yang–Baxter equation, as in reference 66.

Remark 8.9. There are many more complicated triple systems such as the Freudenthal–Kantor triple systems, studied by many authors,[83–90] in connection with explicit constructions of exceptional Lie algebras. Also, it may be worth while remarking that Bars and Günaydin[91] have attempted to utilize triple systems for their work on the construction of possible sub-constituent blocks of quarks and leptons. Other notable works are those by Günaydin,[92] and by Truini and Biedenharn,[93] who utilized the Jordan triple system and Jordan-pair system to generalize the so-called quantum propositional calculus. Finally, more recently, Günaydin and co-workers have published several works for the construction of super-conformal algebras in terms of some triple systems.[94,95] Also, Svinolupov (ref. 95a) has utilized Jordan algebra and the Jordan-pair system for multi-component Korteweg–de Vries and non-linear Schrödinger equations.

9

Non-associative gauge theory

9.1 Digression of Yang–Mills gauge theory

Although we gave a simple account of the Yang–Mills gauge theory in Section 3.4, we will now further elaborate on the theory, so as to pinpoint the difficulties of formulating non-associative Yang–Mills fields.

Let L be a Lie algebra with multiplication table

$$[t_a, t_b] = f_{ab}^c \, t_c \, . \tag{9.1}$$

Then the Yang–Mills gauge field is an L-valued vector field,

$$A_\mu(x) = A_\mu^a(x) \, t_a \, . \tag{9.2}$$

The field strength (or curvature) tensor $F_{\mu\nu}$ is given by

$$F_{\mu\nu} = \partial_\mu A_\nu - \partial_\nu A_\mu + [A_\mu, A_\nu] = \left(\partial_\mu A_\nu^c - \partial_\nu A_\mu^c + f_{ab}^c A_\mu^a A_\nu^b \right) t_c \, . \tag{9.3}$$

Let

$$U(x) = \exp \{ w^a(x) t_a \} \ \epsilon \ G \, , \tag{9.4}$$

which is a coordinate-dependent element of the group G, generated by the Lie algebra L. Now the local gauge transformation for the Yang–Mills field is the transformation

$$A_\mu(x) \longrightarrow A_\mu'(x) = U^{-1}(x) A_\mu(x) U(x) + U^{-1}(x) \partial_\mu U(x) \, , \tag{9.5}$$

under which $F_{\mu\nu}$ transforms covariantly as

$$F_{\mu\nu}(x) \longrightarrow F_{\mu\nu}'(x) = U^{-1}(x) F_{\mu\nu}(x) U(x) \, , \tag{9.6}$$

without the inhomogeneous term as in Eq. (9.5). We can verify the fact

115

that both $A'_\mu(x)$ and $F'_{\mu\nu}(x)$ are L-valued functions, just as $A_\mu(x)$ and $F_{\mu\nu}(x)$, so that we can write

$$A'_\mu(x) = A'^a_\mu(x)t_a \ ,$$
$$F'_{\mu\nu}(x) = F'^a_{\mu\nu}(x)t_a \ . \qquad (9.7)$$

However, the relation between $A^a_\mu(x)$ and $A'^a_\mu(x)$ is, in general, very complicated except for the case of the infinitesimal gauge transformations, in which the $w^a(x)$s in Eq. (9.4) are infinitesimal. Then writing

$$\delta w = w^a(x)t_a = \text{ infinitesimal } \epsilon \ L \ , \qquad (9.8)$$

the relations Eqs. (9.5) and (9.6) are rewritten as

$$\delta A_\mu(x) = A'_\mu(x) - A_\mu(x) = \left[A_\mu(x), \delta w(x)\right] + \partial_\mu \delta w(x) \ , \qquad (9.9a)$$
$$\delta F_{\mu\nu}(x) = F'_{\mu\nu}(x) - F_{\mu\nu}(x) = \left[F_{\mu\nu}(x), \delta w(x)\right] \ . \qquad (9.9b)$$

One important property of the Lie algebra is that the validity of Eqs. (9.9) conversely implies that of Eqs. (9.5) and (9.6) for a finite gauge transformation $U(x)$. Hence, we need consider only infinitesimal gauge transformations. However, the same fact does *not*, in general, apply to other non-associative algebras. Indeed, as we can easily verify, Eqs. (9.9) would hold valid, not only for any Lie algebra, but also for any non-associative algebra. However, this does *not* guarantee the validity of Eqs. (9.5) and (9.6) for the latter.

To be definite, suppose now that Eq. (9.1) is the multiplication table of a non-associative algebra A which is not necessarily a Lie algebra. Then we immediately encounter the problem of how to define $U^{-1}A_\mu U$, since we have

$$\left(U^{-1}A_\mu\right) U \neq U^{-1}\left(A_\mu U\right) \qquad (9.10)$$

in general. If the underlying algebra A is alternative, we can overcome this difficulty by Artin's theorem, which was stated in Section 2.6. However, in order to obtain the transformation law Eq. (9.6) from Eq. (9.5), we need to proceed further using equations such as

$$\left\{U^{-1}\left(A_\mu U\right)\right\}\left\{\left(U^{-1}A_\nu\right)U\right\} = U^{-1}\left\{\left(A_\mu U\right)\left(U^{-1}A_\nu\right)\right\}U$$
$$= U^{-1}\left\{A_\mu(UU^{-1})A_\nu\right\}U$$
$$= U^{-1}\left\{A_\mu A_\nu\right\}U \ , \qquad (9.11)$$

none of which can be proved, even for the alternative algebra, not to mention more complicated non-associative algebras. For the Lie algebra the equalities in Eq. (9.11) are guaranteed by the associativity of the so-called universal enveloping algebra of L, since the products may be interpreted as the associative product in the enveloping algebra. These facts imply that a straightforward generalization of the Yang–Mills gauge theory, for a general non-associative algebra is *not* feasible in general.

In order to discover how we can generalize the theory, let us briefly return to the standard Yang–Mills case, where the underlying algebra is a Lie algebra. It is often convenient to use the terminology of differential geometry. Let $\mathrm{d}x^\mu$ be anti-commuting 1-forms, that is,

$$\mathrm{d}x^\mu \wedge \mathrm{d}x^\nu = -\mathrm{d}x^\nu \wedge \mathrm{d}x^\mu , \qquad (9.12)$$

although they form an associative algebra otherwise. Let Λ_p $(p \geq 0)$ be the vector space consisting of L-valued p-forms with $\Lambda_0 = L$, for $p = 0$. The boundary operator

$$\mathrm{d} \ : \ \Lambda_p \longrightarrow \Lambda_{p+1} \qquad (9.13)$$

satisfies

(i) $\mathrm{dd} = 0$, $\qquad\qquad (9.14a)$

(ii) $\mathrm{d}(\omega_p \omega_q) = (\mathrm{d}\omega_p)\omega_q + (-1)^p \omega_p \mathrm{d}\omega_q$, $\qquad (9.14b)$

for any $\omega_p \in \Lambda_p$ and $\omega_q \in \Lambda_q$. Here and hereafter, we write $\omega_p \ \wedge \ \omega_q$ simply as $\omega_p \omega_q$, in accordance with the usual convention whenever there is no confusion. We now introduce the L-valued 1-form $\omega \in \Lambda_1$ by

$$\omega = A_\mu(x)\mathrm{d}x^\mu \qquad (9.15)$$

and the L-valued 2-form $R \in \Lambda_2$ by

$$R = \mathrm{d}\omega + \omega\omega , \qquad (9.16)$$

which can be rewritten as

$$R = \frac{1}{2} F_{\mu\nu}\mathrm{d}x^\mu \ \wedge \ \mathrm{d}x^\nu \qquad (9.17)$$

in terms of $F_{\mu\nu}$ of Eq. (9.3). Operating d on both sides of Eq. (9.16) and using Eqs. (9.14), we are led to the Bianchi identity

$$\mathrm{d}R = [R, \omega] . \qquad (9.18)$$

After these preparations, the local gauge transformation, Eq. (9.5), can be rewritten as

$$\omega \to \omega' = U^{-1} \omega U + \xi , \qquad (9.19a)$$

where

$$\xi = U^{-1} \mathrm{d} U , \qquad (9.19b)$$

which satisfies the Maurer–Cartan relation

$$d\xi + \xi\xi = 0 \ . \tag{9.20}$$

The curvature 2-form R transforms covariantly as

$$R \rightarrow R' = U^{-1} R U \ . \tag{9.21}$$

In order to eliminate any reference to the associative character of the operator U, we introduce $g \in g\ell(L)$, that is,

$$g \ : \ L \rightarrow L$$

by

$$g \, t_a = U^{-1} \, t_a \, U \ . \tag{9.22}$$

Then, g defines an automorphism (called the inner automorphism). That is, it satisfies

$$g[t_a, t_b] = [gt_a, gt_b] \ . \tag{9.23}$$

We can extend the action of g in a natural fashion to the space Λ_p so that we can also regard it as

$$g \ : \ \Lambda_p \rightarrow \Lambda_p$$

which we denote $g \in g\ell(\Lambda_p)$. Here $g\ell(V)$ stands for the space of the general linear transformations in a vector space V, as before. Eqs. (9.19a) and (9.21) are rewritten, finally, as

$$\omega \rightarrow \omega' = g\omega + \xi \ , \tag{9.24a}$$
$$R \rightarrow R' = gR \ , \tag{9.24b}$$

for $g \in g\ell(\Lambda_p)$ and $\xi \in \Lambda_1$. In this form, the notion of non-associative gauge theory can be formulated, as we see below.

9.2 Non-associative gauge theory

Let A be a non-associative algebra, which is not necessarily a Lie algebra. In Section 3.5, we defined the notion of the derivation Lie algebra of A. Let $D \in g\ell(A)$ be a linear mapping of A into itself, satisfying

$$D(uv) = (Du)v + u(Dv) \tag{9.25}$$

for $u, v \in A$. We also noted that a set consisting of all derivations forms a Lie algebra, which we denoted $L_A (\equiv \mathrm{Der}A)$. If we exponentiate D by

$$g = \exp D \ , \tag{9.26}$$

it is an element of the Lie group G_A, generated by the Lie algebra L_A. Moreover, it satisfies

$$g(uv) = (gu)(gv) . \tag{9.27}$$

That is, g defines an automorphism of the algebra A. Our main idea in constructing a gauge theory based upon A is to utilize the Lie group G_A as the transformation group of the gauge transformation. In order to explain our construction, let us suppose for a moment that we introduce A-valued differential forms, as in the preceding section, with the boundary operator d. As we will see below, we can actually generalize it further. We may introduce the A-valued 1-form $\omega \in \Lambda_1$, as in Eq. (9.15), with the resulting A-valued curvature 2-form $R \in \Lambda_2$ by Eq. (9.16), that is,

$$R = d\omega + \omega\omega . \tag{9.28}$$

Now consider an inhomogeneous transformation

$$\omega \to \omega' = g\omega + \xi , \tag{9.29}$$

just as in Eq. (9.24a), for some $g \in g\ell(\Lambda_1)$ and $\xi \in \Lambda_1$. Assume that they satisfy conditions

(i) $g(\omega\omega) = (g\omega)(g\omega) ,$ $\qquad$ (9.30)

(ii) $d\xi + \xi\xi = 0 ,$ $\qquad$ (9.31)

(iii) $d(g\omega) = g(d\omega) - \xi(g\omega) - (g\omega)\xi .$ $\qquad$ (9.32)

It is then clear that R given by Eq. (9.28) transforms covariantly as

$$R \to R' = d\omega' + \omega'\omega' = gR , \tag{9.33}$$

in conformity with Eq. (9.24b). Actually, we see that we need not assume relations such as Eqs. (9.14) to obtain the covariant transformation law. In particular, we may relax the constraints for d to be a boundary operator. Regardless, the first condition, Eq. (9.30), is automatically satisfied, provided that g is an automorphism of the algebra A, that is, $g \in \text{Aut } A$. We will assume this to be true hereafter. Condition (9.32) can be rewritten in a more compact form as follows. Let us introduce

$$dg \; : \; \Lambda_p \to \Lambda_{p+1} \tag{9.34}$$

by

$$d(g\theta) = (dg)\theta + g(d\theta) \tag{9.35}$$

for $\theta \in \Lambda_p$. Then Eq. (9.32) is rewritten as

$$dg = -L_\xi \, g - R_\xi \, g , \tag{9.36}$$

where L_ξ and R_ξ are left and right multiplication operations, $\Lambda_p \to \Lambda_{p+1}$, defined by

$$L_\xi \theta = \xi \theta \,, \quad R_\xi \theta = \theta \xi \,. \tag{9.37}$$

Let (g_1, ξ_1) and (g_2, ξ_2) be two pairs satisfying Eqs. (9.30)–(9.32). We can verify that they form a group with the composition law

$$(g_1, \xi_1) \circ (g_2, \xi_2) = (g_1 g_2, \xi_1 + g_1 \xi_2) \,, \tag{9.38}$$

where the identity E and the inverse $(g, \xi)^{-1}$ are given by

$$E = (I_d, 0) \tag{9.39a}$$
$$(g, \xi)^{-1} = (g^{-1}, -g^{-1} \xi) \,, \tag{9.39b}$$

where I_d is the identity mapping. Since the pair (g, ξ) forms a Lie group, it really suffices for us to restrict our considerations to infinitesimal transformations for studies of Eqs. (9.30)–(9.32). Let ϵ be an infinitesimally small constant, and let D be a derivation of A. Setting

$$g = \exp(\epsilon D) = I_d + \epsilon\, D + O(\epsilon^2) \,, \tag{9.40a}$$
$$\xi = \epsilon\, \mathrm{d}\, k + O(\epsilon^2) \,, \tag{9.40b}$$

for some $k \in A$ and assuming $\mathrm{dd} = 0$, we see that Eqs. (9.30) and (9.31) are automatically satisfied, while Eq. (9.32) reduces to the validity of

$$\mathrm{d}(D\omega) = D(\mathrm{d}\omega) - (\mathrm{d}k)\omega - \omega(\mathrm{d}k) \,. \tag{9.41}$$

Therefore, we only need to find a pair (D, k) for $k \in A$ and $D \in \mathrm{Der}\, A$, satisfying Eq. (9.41), if we have $\mathrm{dd} = 0$. In case we don't have $\mathrm{dd} = 0$, we have to replace $\mathrm{d}k$ by $\xi_0 \in \Lambda_1$ satisfying $\mathrm{d}\, \xi_0 = 0$ in both Eqs. (9.40b) and (9.41).

Next, we will show that when A is identified with a Lie algebra L, the present result reproduces the standard formulation given in Eq. (9.5). For any $u \in L$, we may then identify

$$D = -ad\, u$$

where $ad\, u$ is the adjoint operation defined by

$$(ad\, u)v = [u, v] \,.$$

Note that $ad\, u \in \mathrm{Der}\, L$ because of the Jacobi identity. Then we can readily identify $k = u$, which satisfies Eq. (9.40). Integrating these, we obtain

$$g = \exp\left(-ad\ u\right) = \sum_{n=0}^{\infty} \frac{1}{n!} \left(-ad\ u\right)^n ,$$

$$\xi = \sum_{n=0}^{\infty} \frac{1}{(n+1)!} \left(-ad\ u\right)^n du ,$$

from which we can easily verify (with $U = \exp\ u$)

$$gv = U^{-1}\ v\ U ,$$

$$\xi = U^{-1}\ d\ U$$

for any $v \in L$, reproducing the results stated in Section 9.1.

Returning to the general case, we note that we require the existence of a symmetric bi-linear non-degenerate form $< u|v >$ in order to construct the Lagrangian of the theory. We also require the invariance condition

$$< gu|gv > = < u|v > \tag{9.42}$$

for $u,\ v \in A$, so that the Yang–Mills Lagrangian

$$L_0 = \frac{1}{4}\ < F_{\mu\nu}|F^{\mu\nu} > \tag{9.43}$$

is invariant under the local gauge transformation. Moreover, suppose that $< u|v >$ satisfies the additional condition

$$< [u,v]|w > = < u|[v,w] > \tag{9.44}$$

for any $u,\ v,\ w \in A$. If the boundary operator d satisfies Eqs. (9.14), the 4-form $< R|R >$ can easily be shown to satisfy the closedness property

$$d < R|R > = 0 , \tag{9.45}$$

as follows. Using the Bianchi identity, Eq. (9.18), we calculate

$$d < R|R > \ = \ < dR|R > \ + \ < R|dR >$$
$$= \ < [R,\omega]|R > \ + \ < R|[R,\omega] > \ = 0$$

by Eq. (9.44), proving Eq. (9.45). We actually obtain

$$< R|R > = \ dW , \tag{9.46a}$$

$$W = < \omega|d\omega > \ + \frac{2}{3}\ < \omega|\omega\omega > \tag{9.46b}$$

in terms of the Chern–Simon form W.

If A is the Lie algebra, which is assumed, further, to be semi-simple, then the symmetric bi-linear non-degenerate form $< u|v >$ is constructed as the Killing form of L given by

$$g_{ab} = < t_a|t_b > = = \mathrm{Tr}\ (\rho(t_a)\rho(t_b)) \tag{9.47}$$

for any non-trivial representation matrix $\rho(t_a)$ of L which also satisfies Eq. (9.44).

Remark 9.1. The details of the present results of this section can be found in reference 96. At present it is not clear whether we can always find a solution of g and ζ for a given non-associative algebra, except for the case of the Lie algebra. This problem will be investigated elsewhere.

9.3 Einstein's relativity as a non-associative Chern–Simon theory

The theory in Section 9.2 has been presented in the language of differential forms as an illustration. However, this is really not necessary, as we will see below. Before going into the details, we reserve the symbol x or x^μ hereafter for the space-time coordinate of a differential manifold M with Dim $M = n$ in order to avoid possible confusion. We also designate by $F(x)$ a set of all smooth functions in M. Now let A be a non-associative algebra, and consider the direct product space

$$\Lambda = A \otimes F(x) , \tag{9.48}$$

which replaces the role of the Λ_ps in the preceding section. Note that Λ is a set of all A-valued smooth functions in M. Further, we reinterpret d to be an A-valued differential operator in Λ, which may not necessarily satisfy Eq. (9.14), that is, dd $= 0$. Also, let us write d and g more explicitly as $d(x)$ and $g(x)$ whenever we wish to exhibit their coordinate dependences.

For any A-valued function $\omega(x)\ \epsilon\ \Lambda$, we introduce the A-valued curvature function $R(x)\ \epsilon\ \Lambda$ by

$$R(x) = \mathrm{d}(x)\omega(x) + \omega(x)\omega(x) . \tag{9.49}$$

To be definite, we will assume hereafter that $\mathrm{d}(x)$ has the form of

$$\mathrm{d}(x) = \beta^\mu \partial_\mu - P , \tag{9.50}$$

where β^μ and P are coordinate-independent elements of $g\ell(A)$.

Consider now the general coordinate transformation

$$x^\mu \to x'^\mu = x'^\mu(x) \quad (\mu = 1, 2, \ldots, n) , \tag{9.51}$$

under which $\omega(x)$ is assumed to change by

$$\omega(x) \rightarrow \omega'(x') = S(x)\omega(x) + \xi(x) \qquad (9.52)$$

for some $S(x) \,\epsilon\, g\ell(\Lambda)$ and $\xi(x) \,\epsilon\, \Lambda$. We have now changed the notation $g = g(x)$ of the preceding section into $S(x)$ in order to avoid confusion with the Riemannian metric tensor $g_{\mu\nu}(x)$, to be introduced shortly. The corresponding transformation law for $R(x)$ is then given by

$$R(x) \rightarrow R'(x') = \mathrm{d}(x')\omega'(x') + \omega'(x')\omega'(x') \,. \qquad (9.53)$$

We now require the validity of the covariance, that is,

$$R'(x') = S(x)R(x) \qquad (9.54)$$

without any inhomogeneous terms depending upon $\xi(x)$. This can be satisfied, provided that we have

(i) $\quad S(x)(\omega(x)\omega(x)) = (S(x)\omega(x))(S(x)\omega(x)) \,, \qquad (9.55a)$

(ii) $\quad \mathrm{d}(x')\xi(x) + \xi(x)\xi(x) = 0 \,, \qquad (9.55b)$

(iii) $\quad \mathrm{d}(x')(S(x)\omega(x)) - S(x)(\mathrm{d}(x)\omega(x))$
$\quad\quad +\xi(x)(S(x)\omega(x)) + (S(x)\omega(x))\xi(x) = 0 \,. \qquad (9.55c)$

Comparing these with Eqs. (9.30)–(9.32), we now see the appearance of both $\mathrm{d}(x)$ and $\mathrm{d}(x')$ because of the coordinate transformation. However, Eq. (9.55a) is unchanged, with $S(x)$ again being a local automorphism of A.

In order to recast Einstein's general relativity into the framework of the present theory, let

$$g^{\mu\nu}(x) = g^{\nu\mu}(x) \,, \quad \Gamma^\lambda_{\mu\nu}(x) = \Gamma^\lambda_{\nu\mu}(x) \,, \qquad (9.56)$$

be the metric tensor and Christoffel's symbol respectively. Consider the Palatini action[97]

$$W = \int \mathrm{d}^n x \sqrt{g(x)} \; g^{\mu\nu}(x) \{ \partial_\nu \Gamma_\mu(x) - \partial_\lambda \Gamma^\lambda_{\mu\nu}(x)$$
$$- \Gamma^\lambda_{\mu\nu}(x)\Gamma_\lambda(x) + \Gamma^\alpha_{\mu\beta}(x)\Gamma^\beta_{\alpha\nu}(x) \} \,. \qquad (9.57)$$

It is well-known that the action principle $\delta W = 0$ for independent variations of $g^{\mu\nu}(x)$, $\Gamma^\lambda_{\mu\nu}(x)$, and $\Gamma_\mu(x)$ leads to the Einstein equation of motion with

$$\Gamma_\mu(x) = \Gamma^\lambda_{\mu\lambda}(x) \,, \qquad (9.58)$$

provided that the space-time dimension n is larger than 2. Our strategy is to rewrite W in the Chern–Simon form

$$W = \; <\omega|d\omega> \; + \frac{2}{3} \; <\omega|\omega\omega> \;, \qquad (9.59)$$

with

$$<u|v> \; = \; \frac{1}{4} \int d^n x \; (u(x)|v(x)) \qquad (9.60)$$

for some A-valued function ω and for some bi-linear symmetric form $(u|v)$ in A. To this end, suppose that ω is expressed in the form of

$$\omega = \sqrt{g(x)} \; g^{\mu\nu}(x)e_{\mu\nu} + \Gamma^{\lambda}_{\mu\nu}(x)e^{\mu\nu}_{\lambda} + \Gamma_{\mu}(x)e^{\mu} \;, \qquad (9.61)$$

where $e_{\mu\nu}$, $e^{\mu\nu}_{\lambda}$, and e^{μ} are some coordinate-independent elements of a non-associative algebra A, to be specified shortly. However, A must contain other elements in order to obtain a satisfactory result. The minimum choice for A appears to require that it is spanned as

$$A = \{e_{\mu}, e^{\mu}, e_{\mu\nu}, e^{\mu\nu}, e^{\mu\nu}_{\lambda}, e^{\lambda}_{\mu\nu}, e^{\mu}_{\nu}\} \;, \qquad (9.62)$$

satisfying the symmetry condition

$$e_{\mu\nu} = e_{\nu\mu} \;, \quad e^{\mu\nu}_{\lambda} = e^{\nu\mu}_{\lambda} \;, \qquad (9.63)$$

etc. Here, e^{μ}_{ν} could be identified with $\delta^{\mu}_{\nu} I$. Moreover, we introduce the symmetric bi-linear form $(u|v)$ in A by

$$(e_{\mu}|e^{\nu}) = \delta^{\nu}_{\mu} \;, \qquad (9.64a)$$

$$(e_{\mu\nu}|e^{\alpha\beta}) = \delta^{\alpha}_{\mu}\delta^{\beta}_{\nu} + \delta^{\alpha}_{\nu}\delta^{\beta}_{\mu} \;, \qquad (9.64b)$$

$$(e^{\mu\nu}_{\lambda}|e^{\gamma}_{\alpha\beta}) = \delta^{\gamma}_{\lambda}(\delta^{\mu}_{\alpha}\delta^{\nu}_{\beta} + \delta^{\mu}_{\beta}\delta^{\nu}_{\alpha}) \;, \qquad (9.64c)$$

with all other $(u|v)$ being identically zero except possibly for $\left(e^{\mu}_{\nu}|e^{\alpha}_{\beta}\right)$.

After these preparations, the simplest choice for d is to set $P = 0$ in Eq. (9.50) and hence

$$d(x) = \beta^{\lambda}\partial_{\lambda} \;, \qquad (9.65)$$

where $\beta^{\lambda} \; \epsilon \; g\ell(A)$ acts upon $e^{\mu}, e_{\mu\nu}$, and $e^{\mu\nu}_{\alpha}$ as

$$\beta^{\lambda}e^{\mu} = e^{\lambda\mu} \;, \qquad (9.66a)$$

$$\beta^{\lambda}e_{\mu\nu} = e^{\lambda}_{\mu\nu} - \delta^{\lambda}_{\mu}e_{\nu} - \delta^{\lambda}_{\nu}e_{\mu} \;, \qquad (9.66b)$$

$$\beta^{\lambda}e^{\mu\nu}_{\alpha} = -\delta^{\lambda}_{\alpha}e^{\mu\nu} \;. \qquad (9.66c)$$

Then we can readily calculate

$$<\omega|d\omega> \; = \int d^n x \; \sqrt{g(x)} \; g^{\mu\nu}(x) \left\{\partial_{\nu}\Gamma_{\mu}(x) - \partial_{\lambda}\Gamma^{\lambda}_{\mu\nu}(x)\right\} \qquad (9.67)$$

where we discarded the surface integral resulting from integration in part. As we will note in Remark 9.2 at the end of this section, we need not specify the action of β^λ upon other elements e_μ, $e^{\mu\nu}$, and $e_{\mu\nu}^\alpha$ in order to obtain the desired result, Eq. (9.67), which represents the first two terms in the right-hand side of Eq. (9.57). To recover the rest of the terms in Eq. (9.57), we now assume the following partial multiplication table for A:

$$e^\mu e^\nu = e_{\mu\nu} e_{\alpha\beta} = 0 \,, \tag{9.68a}$$

$$e_{\mu\nu} e^\lambda = e^\lambda e_{\mu\nu} = -\frac{1}{2} \, e_{\mu\nu}^\lambda \,, \tag{9.68b}$$

$$e_\lambda^{\mu\nu} e^\alpha = e^\alpha e_\lambda^{\mu\nu} = -\frac{1}{2} \, \delta_\lambda^\alpha e^{\mu\nu} \,, \tag{9.68c}$$

$$
\begin{aligned}
e_\lambda^{\mu\nu} e_\gamma^{\alpha\beta} = \frac{1}{4} \, \big(&\delta_\gamma^\mu \delta_\lambda^\alpha e^{\nu\beta} + \delta_\gamma^\mu \delta_\lambda^\beta e^{\nu\alpha} \\
&+ \delta_\gamma^\nu \delta_\lambda^\alpha e^{\mu\beta} + \delta_\gamma^\nu \delta_\lambda^\beta e^{\mu\alpha} \big) \,,
\end{aligned}
\tag{9.68d}
$$

$$
\begin{aligned}
e_\lambda^{\mu\nu} e_{\alpha\beta} = e_{\alpha\beta} e_\lambda^{\mu\nu} = &-\frac{1}{2} \, \big(\delta_\alpha^\mu \delta_\beta^\nu + \delta_\beta^\mu \delta_\alpha^\nu \big) e_\lambda \\
&+ \frac{1}{4} \, \big\{ \delta_\alpha^\mu e_{\lambda\beta}^\nu + \delta_\alpha^\nu e_{\lambda\beta}^\mu + \delta_\beta^\mu e_{\lambda\alpha}^\nu + \delta_\beta^\nu e_{\lambda\alpha}^\mu \big\} \,,
\end{aligned}
\tag{9.68e}
$$

which are sufficient to give

$$< \omega | \omega \omega > = \frac{3}{2} \int d^n x \sqrt{g(x)} g^{\mu\nu}(x) \left\{ \Gamma_{\mu\beta}^\alpha(x) \Gamma_{\alpha\nu}^\beta(x) - \Gamma_{\mu\nu}^\lambda(x) \Gamma_\lambda(x) \right\}. \tag{9.69}$$

Therefore, we have succeeded in rewriting the Palatini action, Eq. (9.57), in the Chern–Simon form, Eq. (9.59). Again, we need not specify the rest of the multiplication table of A, such as for the products $e_\mu e^\nu$, $e_{\alpha\beta} e_{\mu\nu}^\lambda$, etc. (See Remark 9.2 at the end of this section.) In this connection, we note that Eqs. (9.68) are consistent with A being commutative, that is,

$$uv = vu \tag{9.70}$$

for u, $v \in A$. Moreover, we also note that Eqs. (9.66) are compatible with the relation

$$(u | \beta^\lambda v) + (\beta^\lambda u | v) = 0 \tag{9.71a}$$

and hence

$$< u | dv > \, = \, < du | v > \,, \tag{9.71b}$$

as well as

$$< uv | w > \, = \, < u | vw > \,, \tag{9.71c}$$

for any u, v, $w \in \Lambda$. Then when we vary ω into $\omega + \delta\omega$, the variational principle $\delta W = 0$ for the Chern–Simon Lagrangian, Eq. (9.59), leads to the equation of motion

$$R = \ d\omega + \omega\omega = 0 \,, \tag{9.72}$$

by Eqs. (9.71b) and (9.71c). On the other hand, we calculate R by a direct computation to be

$$R = \{\partial_\nu \Gamma_\mu - \partial_\lambda \Gamma^\lambda_{\mu\nu} - \Gamma^\lambda_{\mu\nu}\Gamma_\lambda + \Gamma^\lambda_{\beta\nu}\Gamma^\beta_{\lambda\mu}\}e^{\mu\nu}$$
$$+ \{\partial_\lambda(\sqrt{g}\, g^{\mu\nu}) + 2\sqrt{g}\, g^{\alpha\nu}\Gamma^\mu_{\alpha\lambda} - \sqrt{g}\, g^{\mu\nu}\Gamma_\lambda\}e^\lambda_{\mu\nu}$$
$$- 2\{\sqrt{g}\, g^{\mu\nu}\Gamma^\alpha_{\mu\nu} + \partial_\lambda(\sqrt{g}\, g^{\alpha\lambda})\}e_\alpha \tag{9.73}$$

from Eqs. (9.65), (9.66), and (9.68). Therefore (by setting coefficients of $e^{\mu\nu}$, $e^\lambda_{\mu\nu}$, and e_α to zero), $R = 0$ leads to Einstein's equation

$$R_{\mu\nu} = \partial_\lambda \Gamma^\lambda_{\mu\nu} - \frac{1}{2}\left(\partial_\mu \Gamma_\nu + \partial_\nu \Gamma_\mu\right) + \Gamma^\lambda_{\mu\nu}\Gamma_\lambda - \Gamma^\beta_{\alpha\nu}\Gamma^\alpha_{\beta\mu} = 0 \,, \tag{9.74a}$$

as well as to two other familiar relations

$$\Gamma_\lambda = \Gamma^\alpha_{\lambda\alpha} = \partial_\lambda \log \sqrt{g} \,, \tag{9.74b}$$

$$\Gamma^\lambda_{\mu\nu} = \frac{1}{2}\, g^{\lambda\alpha}\left(\partial_\mu g_{\alpha\nu} + \partial_\nu g_{\alpha\mu} - \partial_\alpha g_{\mu\nu}\right) \,, \tag{9.74c}$$

provided that the space-time dimension $n = \ \text{Dim}\, M$ is greater than 2.

We can now find explicit expressions for which $S(x)$ and $\xi(x)$ satisfy Eqs. (9.55). First, the action of $S(x) \in g\ell(\Lambda)$ on elements of A is defined by

$$S(x)e_{\mu\nu} = \left|\frac{\partial x}{\partial x'}\right| \frac{\partial x'^\alpha}{\partial x^\mu} \frac{\partial x'^\beta}{\partial x^\nu}\, e_{\alpha\beta} \,, \tag{9.75a}$$

$$S(x)e^\mu = \frac{\partial x^\mu}{\partial x'^\lambda}\, e^\lambda \,, \tag{9.75b}$$

$$S(x)e^{\mu\nu}_\lambda = \frac{\partial x'^\gamma}{\partial x^\lambda} \frac{\partial x^\mu}{\partial x'^\alpha} \frac{\partial x^\nu}{\partial x'^\beta}\, e^{\alpha\beta}_\gamma \,, \tag{9.75c}$$

$$S(x)e^\lambda_{\mu\nu} = \left|\frac{\partial x}{\partial x'}\right| \frac{\partial x'^\alpha}{\partial x^\mu} \frac{\partial x'^\beta}{\partial x^\nu} \frac{\partial x^\lambda}{\partial x'^\gamma}\, e^\gamma_{\alpha\beta} \,, \tag{9.75d}$$

$$S(x)e^{\mu\nu} = \frac{\partial x^\mu}{\partial x'^\alpha} \frac{\partial x^\nu}{\partial x'^\beta}\, e^{\alpha\beta} \,, \tag{9.75e}$$

$$S(x)e_\mu = \left|\frac{\partial x}{\partial x'}\right| \frac{\partial x'^\lambda}{\partial x^\mu}\, e_\lambda \,, \tag{9.75f}$$

where we have set

$$\left|\frac{\partial x}{\partial x'}\right| = \det \left|\frac{\partial x^\nu}{\partial x'^\mu}\right| . \tag{9.76}$$

Then we can readily verify the validity of

$$S(x)(\omega(x)\omega(x)) = (S(x)\omega(x))(S(x)\omega(x)) . \tag{9.77}$$

Moreover, setting

$$\xi(x) = -\frac{\partial x^\alpha}{\partial x'^\mu}\frac{\partial x^\beta}{\partial x'^\nu}\frac{\partial^2 x'^\lambda}{\partial x^\alpha \partial x^\beta} e_\lambda^{\mu\nu}$$

$$-\frac{\partial x^\alpha}{\partial x'^\mu}\frac{\partial x^\beta}{\partial x'^\lambda}\frac{\partial^2 x'^\lambda}{\partial x^\alpha \partial x^\beta} e^\mu , \tag{9.78}$$

we can see that the transformation law, Eq. (9.52), that is,

$$\omega(x) \rightarrow \omega'(x) = S(x)\omega(x) + \xi(x) ,$$

reproduces the familiar formula:

$$\Gamma_{\mu\nu}^\lambda(x) \rightarrow \Gamma_{\mu\nu}'^\lambda(x') = \frac{\partial x'^\lambda}{\partial x^\gamma}\frac{\partial x^\alpha}{\partial x'^\mu}\frac{\partial x^\beta}{\partial x'^\nu}\Gamma_{\alpha\beta}^\gamma(x) - \frac{\partial x^\alpha}{\partial x'^\mu}\frac{\partial x^\beta}{\partial x'^\nu}\frac{\partial^2 x'^\lambda}{\partial x^\alpha \partial x^\beta} , \tag{9.79a}$$

as well as

$$\Gamma_\mu(x) \rightarrow \Gamma_\mu'(x') = \frac{\partial x^\lambda}{\partial x'^\mu}\Gamma_\lambda(x) - \frac{\partial x^\alpha}{\partial x'^\mu}\frac{\partial x^\beta}{\partial x'^\lambda}\frac{\partial^2 x'^\lambda}{\partial x^\alpha \partial x^\beta} . \tag{9.79b}$$

We can further verify the validity of Eqs. (9.55). We remark that Eqs. (9.75) give

$$(S(x)u(x)|S(x)v(x)) = \left|\frac{\partial x}{\partial x'}\right|(u(x)|v(x)) \tag{9.80}$$

for $u(x)$, $v(x) \in \Lambda$, so that we have

$$\int d^n x' (S(x)u(x)|S(x)v(x)) = \int d^n x (u(x)|v(x)) . \tag{9.81}$$

Next, let us restrict ourselves to the infinitesimal coordinate transformation

$$x^\mu \rightarrow x'^\mu = x^\mu + \varepsilon \; \phi^\mu(x) + O(\varepsilon^2) \tag{9.82}$$

for some infinitesimal constant ε with a smooth function $\phi^\mu(x)$. Then we have

$$\xi(x) = -\varepsilon \left(\frac{\partial^2 \phi^\lambda}{\partial x^\mu \partial x^\nu} e_\lambda^{\mu\nu} + \frac{\partial^2 \phi^\lambda}{\partial x^\mu \partial x^\lambda} e^\mu\right) + O(\varepsilon^2) . \tag{9.83}$$

Similarly, we expand $S(x)$ as

$$S(x) = I + \varepsilon \frac{\partial \phi^v}{\partial x^\mu} X_v^\mu + O(\varepsilon^2) , \tag{9.84}$$

where $X_v^\mu \in g\ell(A)$ is coordinate-independent. Now the action of X_v^μ on elements of A can readily be found to be

$$X_v^\mu e^\lambda = -\delta_v^\lambda e^\mu , \tag{9.85a}$$

$$X_v^\mu e_{\alpha\beta} = \delta_\alpha^\mu e_{v\beta} + \delta_\beta^\mu e_{\alpha v} - \delta_v^\mu e_{\alpha\beta} , \tag{9.85b}$$

$$X_v^\mu e_\lambda^{\alpha\beta} = \delta_\lambda^\mu e_v^{\alpha\beta} - \delta_v^\alpha e_\lambda^{\mu\beta} - \delta_v^\beta e_\lambda^{\alpha\mu} , \tag{9.85c}$$

$$X_v^\mu e_\lambda = \delta_\lambda^\mu e_v - \delta_v^\mu e_\lambda , \tag{9.85d}$$

$$X_v^\mu e^{\alpha\beta} = -\delta_v^\alpha e^{\mu\beta} - \delta_v^\beta e^{\alpha\mu} , \tag{9.85e}$$

$$X_v^\mu e_{\alpha\beta}^\lambda = \delta_\alpha^\mu e_{v\beta}^\lambda + \delta_\beta^\mu e_{\alpha v}^\lambda - \delta_v^\lambda e_{\alpha\beta}^\mu - \delta_v^\mu e_{\alpha\beta}^\lambda . \tag{9.85f}$$

In particular, the X_v^μs satisfy the $g\ell(n)$ Lie algebra relation

$$[X_v^\mu, X_\beta^\alpha] = \delta_\beta^\mu X_v^\alpha - \delta_v^\alpha X_\beta^\mu , \tag{9.86}$$

implying that the derivation algebra of A is the Lie algebra $g\ell(n)$.

Remark 9.2. An inspection of the present theory shows that we can relax our assumption as follows. Set

$$B = \{e_{\mu v}, e_\lambda^{\mu v}, e^\lambda\} , \tag{9.87a}$$

and

$$B^* = \{e^{\mu v}, e_{\mu v}^\lambda, e_\lambda\} , \tag{9.87b}$$

so that

$$A \supset B \cup B^* . \tag{9.88}$$

We only require the validity of

$$B\,B \subset B^* , \tag{9.89a}$$

and

$$d\,B \subset B^* . \tag{9.89b}$$

In particular, if we choose $dB^* = 0$, then we can satisfy $dd = 0$ and $d(uv) = (du)v + u(dv)$ in B. Moreover, the symmetric bi-linear product $< B^*|B > = < B|B^* >$ must satisfy the conditions

$$< uv|w > \, = \, < u|vw > , \tag{9.90a}$$

$$< u|dv > \, = \, < du|v > , \tag{9.90b}$$

for any u, v, $w \in B$. In particular, we need *not* define the products BB^* or B^*B^* as well as dB^* in order to rewrite the Palatini action into the Chern–Simon form.

Remark 9.3. The Chern–Simon theory for an associative, but non-commutative, algebra has been used already by Witten[98] for the second quantization of the open-string model.

Remark 9.4. It appears that we can generalize the Einstein theory into an octonionic theory.[99] This problem will be studied elsewhere on the basis of the present formulation.

Remark 9.5. The standard Yang–Mills gauge theory can also be recast in Chern–Simon form, with some non-associative algebra. However, we will not go into the details here.

Remark 9.6. We could integrate the infinitesimal coordinate transformation Eq. (9.82) as follows. Let t be a real parameter $0 \leq t \leq 1$ and set

$$X^\mu(x,t) = \exp\left\{t\phi^\lambda(x)\frac{\partial}{\partial x^\lambda}\right\} x^\mu \,, \tag{9.91}$$

which satisfies the initial condition at $t = 0$,

$$X^\mu(x,0) = x^\mu \,, \tag{9.92}$$

as well as

$$\frac{\partial}{\partial t} X^\mu(x,t) = \phi^\lambda(x)\frac{\partial}{\partial x^\lambda} X^\mu(x,t) \,. \tag{9.93}$$

Moreover, for $t = \epsilon$ being infinitesimal, we have

$$X^\mu(x,\epsilon) = x^\mu + \epsilon\phi^\mu(x) + O(\epsilon^2) \,, \tag{9.94}$$

which is Eq. (9.82). Hence, the desired finite coordinate transformation corresponding to the infinitesimal one in Eq. (9.82) is the solution of the differential equation (9.93) with the initial condition (9.92). Then we may make the identification

$$x'^\mu = X^\mu(x,1) \tag{9.95}$$

at $t = 1$. The interplay between the infinitesimal and finite transformations is made possible, of course, by the Lie-group character of all differentiable coordinate transformations.

10

Concluding remark

In this short treatise, we have given some ideas for applying non-associative algebras to physical problems. However, we have had to omit many other topics related to our subject in order to make our treatment as self-contained as possible. We have not mentioned the important technical method of the Peirce decomposition,[7] which is crucial for determining the structures of many non-associative algebras. Nor have we presented the subject of the Tit construction,[7] or the Freudenthal–Kantor constructions of exceptional Lie algebras.[82–90] Similarly, the representation theory of Jordan algebra[100] might have been addressed. Regarding physics, we have not touched on the matter of quaternionic[101,102] and octonionic[103] Hilbert spaces, and their possible relevances to physics. Nor has the subject of non-associative generalizations of the so-called propositional calculus and related subjects in quantum mechanics been given any attention. The latter problems have been discussed by Günaydin[92] and by Truini and Biedenharn,[93] who utilized the Jordan triple system and the Jordan-pair systems. Moreover, nothing has been said on the possible relevance[104,105] of non-associative algebras for second quantizations of closed string models, in view of the so-called non-associative anomaly. Also, the triangle anomaly in quantum field theory may lead to a violation[106] of the Jacobi identity for the algebra of currents, leading to non-associative physics. A non-associative algebra also occurs[107] in some integrable classical mechanics problem.[95a] Further, a non-associative algebra has been utilized[108] for construction of the B-type vertex operator in the string model. Lastly, we have scarcely mentioned the notion of non-associative super algebras. We simply note that an attempt to find composition super-algebras has been made elsewhere.[109] Moreover, Kac[110] has classified simple Jordan-super algebras, which is of some interest in connection with the study of the Poisson bracket.[60,111]

References

1. L. Novey, *Origins of modern algebras*, Noordhoff Inst. Publ., Leyden (1973).
2. M. L. Tomber, A short history of non-associative algebras, in *Proceedings of the 2nd Workshop on Lie-admissible Formulation, Hadronic Journal*, 2, 1252 (1979).
3. W. R. Hamilton, *Report of the British Association for the Advancement of Science for 1834*, pp. 519–23.
4. H. Freudenthal, *Oktaven, Ausnahmegruppen und Oktavengeometrie*, Math. Inst. Rijksuniversiteit, Holland (1960).
5. M. Günaydin and F. Gürsey, *Jour. Math. Phys.*, **14**, 1651 (1973).
6. S. Okubo and Y. Tosa, *Phys. Rev.*, **D20**, 462 (1979), Erratum **D23**, 1468 (1981).
7. R. D. Schafer, *An introduction to non-associative algebras*, Academic Press, New York and London (1966).
8. M. Günaydin, C. Piron and M. Ruegg, *Comm. Math. Phys.*, **61**, 69 (1978).
9. K. M. Case, *Phys. Rev.*, **97**, 810 (1955).
10. S. Okubo and H. C. Myung, *Jour. Algebra*, **67**, 479 (1980).
11. A. Sudberg, *Jour. Phys. A. Math. Gen.*, **17**, 939 (1984).
12. F. R. Harvey, *Spinors and calibrations*, Academic Press, New York 1990.
13. N. Jacobson, *Rend. Circ. Mat. Palermo*, II **7**, 55 (1958).
14. O. K. Kalashinikov, E. S. Fradkin and E. E. Konstein, *Sov. Jour. Nucl. Phys.*, **29**, 852 (1979).
15. A. M. Polyakov, *Gauge fields and strings*, Harwood Academic Pub., London/New York (1987).
16. A. A. Belavin, A. M. Polyakov, A. S. Schwartz and Yu. S. Tyupkin, *Phys. Lett.*, **59B**, 85 (1975).
17. R. Peccei, *CP violation*, ed. by C. Jarlskog, World Scientific, Singapore, p. 503 (1989).
18. S. Okubo and R. E. Marshak, *Prog. Theor. Phys.*, **87**, 1059 (1992).
19. D. S. Freed and K. K. Uhlenbeck, *Instantons and four-manifolds*, Springer-Verlag, New York/Berlin/Heidelberg/Tokyo (1984).
20. L. Sorgsepp and J. Lõhmus, *Hadronic Jour.*, **4**, 327 (1981).
21. S. Catto and F. Gürsey, *Nuovo Cimento*, **86A**, 201 (1985) and **99A** 685 (1988).
22. H. C. Myung, *Malcev-admissible algebras*, Birkhäuser, Boston (1986).
23. S. Okubo, *Hadronic Jour.*, **1**, 1250, 1383 and 1432 (1978).
24. M. Gell-Mann, *Phys. Rev.*, **125**, 1067 (1962).

25. S. Okubo, *Jour. Math. Phys.*, **32**, 1657 and 1669 (1991), *Mathematica Japonica* (to appear).
26. S. Okubo and J. M. Osborn, *Comm. in Algebra*, **9**, 1233 and 2015 (1981).
27. S. Okubo, *Hadronic Jour.*, **5**, 1564 and 1613 (1982).
28. A. Elduque and H. C. Myung, *Comm. in Algebra*, **21**, 2481 (1993).
29. S. Okubo, *Hadronic Jour.*, **4**, 216 (1981).
30. D. B. Shapiro, quoted in reference 22, p.62.
31. M. L. El-Mallah, *Archiv der Math.*, **49**, 16 (1987) and **51**, 39 (1988).
32. J. Milnor and R. Bott, *Bull. Amer. Math. Soc.*, **64**, 87 (1958).
33. M. Kervair, *Proc. Nat. Acad. Sci.*, **44**, 286 (1958).
34. G. Frobenius, *J. reine u. angew. Math.*, **84**, 59 (1876).
35. A. G. Kurosh, *Lectures on general algebras*, Chelsea, NY (1963).
36. G. M. Benkart and J. M. Osborn, *Am. Jour. Math.*, **103**, 1135 (1979) and *Pacific Jour. Math.*, **96**, 265 (1981).
37. G. M. Benkart, G. M. Britten and J. M. Osborn, *Canad. J. Math.*, **34**, 550 (1982).
38. F. J. Dyson, *Jour. Math. Phys.*, **3**, 140, 157 (1962).
39. F. Gliozzi, J. Sherk and D. Olive, *Nucl. Phys.*, **B122**, 253 (1977).
40. M. B. Green, J. H. Schwarz and E. Witten, *Superstring theory I and II*, Cambridge University Press, London 1987.
41. P. Goddard, W. Nahm, D. I. Olive, H. Ruegg and A. Schwimmer, *Comm. Math. Phys.*, **112**, 385 (1987).
42. G. 'tHooft and M. Veltman, *Nucl. Phys.*, **B44**, 189 (1972).
43. F. B. Little, R. B. Mann, V. Elias and D. G. C. McKeon, *Phys. Rev.*, **D32**, 2707 (1985) and earlier references quoted therein.
44. S. Okubo, *Prog. Theor. Phys. Suppl.*, **86**, 287 (1986).
45. M. D. Rose, *Elementary theory of angular momentum*, Wiley, NY (1957).
46. L. C. Biedenharn and J. D. Louck, *Angular momentum in quantum physics: theory and application*, Addison-Wesley, Reading, MA (1981).
47. G. M. Benkart and J. M. Osborn, *Pac. J. Math.*, see ref. 36.
48. D. E. Littlewood, *The theory of group characters*, Clarendon, Oxford (1940).
49. H. Weyl, *Classical groups*, Princeton University Press, Princeton, NY (1939).
50. S. Okubo, *Alg. Group. Geom.*, **3**, 60 (1986).
51. W. G. McKay and J. Patera, *Tables of dimensions, indices, and branching rules for representations of simple Lie algebras*, Dekker, New York (1981).
52. S. Okubo and J. Patera, *Jour. Math. Phys.*, **24**, 2722 (1983), **25**, 219 (1984).
53. P. Jordan, *Nachr. Ges. Wiss. Göttinger*, 569 (1932) and 209 (1933), *Z. Physik*, **80**, 285 (1933).
54. P. Jordan, J. von Neumann and E. Wigner, *Am. J. of Math.*, **35**, 29 (1934).
55. E. I. Zelmanov, *Siberian Math. Jour.*, **24**, 89 (1983), quoted by K. McCrimman, *Alg. Group and Geom.*, **1**, 1 (1984). The author would like to express his gratitude to Prof. H. C. Myung for informing him of these references.
56. S. Okubo, *Hadronic Jour.*, **4**, 608 (1981), **5**, 1667 (1982) and in *Proceedings of the 3rd Workshop on Current Problems in High Energy Particle Theory*, ed. by R. Casalbuoni, G. Domokos, and S. Kovesi-Domokos, John-Hopkins University Press, p.103 (1979).
57. S. Okubo and H. C. Myung, *Trans. Amer. Math. Soc.*, **264**, 459 (1981).

58. G. M. Benkart and J. M. Osborn, *Jour. Algebra*, **71**, 11 (1981).
59. R. M. Santilli, *Lett. Nuov. Cimento*, **37**, 337 (1983) and **38**, 509 (1983), and earlier references quoted therein.
60. I. L. Kantor, in *Lie theory, differential equations and representation theory*, ed. by V. Hussein, Université de Montréal Press, Montréal (1990).
61. S. Weinberg, *Nucl. Phys. B (Proc. Suppl.)*, **6**, 67 (1989) and University of Texas Report UTTG-30-87/08-89.
62. S. Okubo, *Phys. Rev.*, **C10**, 2045 and 2048 (1974).
62a. J. E. Moyal, *Proc. Cambridge Phil. Soc.*, **45**, 99 (1949), F. Bayen *et al.*, *Ann. of Phys.*, **111**, 61 and 111 (1978).
63. S. Okubo, *Hadronic Jour.*, **1**, 1383, 1432 (1978) and **2**, 39 (1979).
64. N. Jacobson, *Lie algebras*, Interscience, NY (1962).
65. L. J. Paige in *Studies in modern algebra vol. 2*, edited by A. A. Albert, Mathematical Association of America, Prentice Hall, p. 180 (1963).
66. S. Okubo, *Jour. Math. Phys.*, **34**, 3273, 3292 (1993).
67. B. de Wit and H. Nicolai, *Nucl. Phys.*, **B 231**, 506 (1984).
68. F. Gürsey and C. H. Tze, *Phys. Lett.*, **127B**, 191 (1983).
69. *Yang–Baxter equation in integrable systems*, ed. by M. Jimbo, World Scientific, Singapore (1989).
70. *Braid group, knot theory and statistical mechanics*, ed. by C. N. Yang and M. L. Ge, World Scientific, Singapore, (1989).
71. L. H. Kauffman, *Knots and physics*, World Scientific, Singapore (1991).
72. Y. I. Manin, *Quantum groups and non-commutative geometry*, University of Montréal Press, Montréal (1988).
73. A. B. Zamolodchikov and Al. B. Zamolodchikov, *Nucl. Phys.*, **B133**, 525 (1978).
74. H. J. de Vega and H. Nicolai, *Phys. Lett.*, **B244**, 295 (1990).
75. Y. Nambu, *Phys. Rev.*, **D7**, 2405 (1973).
76. W. G. Lister, *Amer. Jour. Math.*, **89**, 787 (1952).
77. K. Yamaguchi, *Jour. Sci. Hiroshima University*, **A21**, 155 (1958).
78. S. Helgason, *Differential geometry and symmetric spaces*, Academic Press, NY (1962).
79. Y. Ohnuki and S. Kamefuchi, *Quantum field theory and parastatistics*, University of Tokyo Press/Springer, Tokyo/Berlin (1982).
80. S. Okubo, *Jour. Math. Phys.*, **35**, 2785 (1994).
81. I. Bialynicki-Birula and P. J. Morrison, *Phys. Lett.*, **A158**, 453 (1991).
82. K. Yamaguchi and H. Asano, *Proc. Jap. Acad.*, **51**, 247 (1972).
83. J. R. Faulker and J. C. Ferrar, *Indago Math.*, **34**, 247 (1972).
84. I. L. Kantor, *Sov. Math. Dokl.*, **14**, 254 (1977).
85. W. Hein, *Trans. Amer. Math. Soc.*, **205**, 79 (1975) and *Math. Ann.*, **213**, 195 (1975).
86. B. N. Allison, *Amer. Jour. Math.*, **98**, 285 (1970).
87. I. Bars and M. Günaydin, *Jour. Math. Phys.*, **20**, 1977 (1979).
88. Y. Kakiichi, *Proc. Jap. Acad.*, **57**, Ser A 276 (1981).
89. K. Yamaguchi, *Bull. Fac. Sch. Ed. Hiroshima University*, **6** (2), 49 (1983).
90. K. Yamaguchi and A. Ono, *Bull. Fac. Sch. Ed. Hiroshima University*, Part II, **7**, 43 (1984).
91. I. Bars and M. Günaydin, *Phys. Rev.*, **D22**, 1403 (1980).
92. M. Günaydin, *Ann. Israel Phys. Soc.*, **3**, 279 (1980).
93. P. Truini and L. C. Biedenharn, *Jour. Math. Phys.*, **23**, 1327 (1982).
94. M. Günaydin, *Phys. Lett.*, **B255**, 46 (1991).

95. M. Günaydin and S. Hyun, *Mod. Phys. Lett.*, **6**, 1733 (1991) and *Nucl. Phys.*, **B373**, 688 (1992).

95a. S. I. Svinolupov, *Comm. Math. Phys.*, **143**, 559 (1992) and *Funktsional'nyi Analiz iEgo Frilozheniya*, **27**, 40 (1993). (The author would like to express his gratitude to Professor O. Smirnov for informing him of these references and for sending him reprints.)

96. S. Okubo, *Alg. Group. Geom.*, **4**, 215 (1987).

97. C. W. Misner, K. S. Thorne and J. A. Wheeler, *Gravitation*, Freeman and Co., San Francisco, (1971).

98. E. Witten, *Nucl. Phys.*, **B268**, 253 (1986).

99. S. Marques and C. G. Oliveira, *Phys. Rev.*, D **36**, 1716 (1987) and *Jour. Math. Phys.*, **26**, 3131 (1985).

100. N. Jacobson, Structure and representations of Jordan algebras, *Amer. Math. Sci. Coll. Pub.*, **39**, Amer. Math. Soc., Providence, RI (1968).

101. e.g. see P. Truini, L. C. Biedenharn and G. Cassinelli, *Hadronic Jour.*, **4**, 981 (1981) and earlier references quoted therein.

102. S. L. Adler, in *From symmetries to strings*, 40 years of Rochester Conferences, ed. by A. Das, World Scientific (1990).

103. e.g. see M. Günaydin, *Jour. Math. Phys.*, **17**, 1875 (1976).

104. G. T. Horowitz and A. Strominger, *Phys. Lett.*, **B185**, 45 (1987).

105. A. Strominger, *Phys. Rev. Lett.*, **58**, 629 (1987), *Nucl. Phys.*, **B294**, 93 (1987).

106. R. Jackiw, *Phys. Rev. Lett.*, **54**, 159 (1985).

107. A. A. Balinskii and S. P. Novikov, *Dokl. Akad. Nauk SSSR*, **283**, 1036 (1985). English transl. *Soviet Math. Dokl.*, **32**, 228 (1985).

108. M. V. Cougo-Pinto, *Jour. Math. Phys.*, **29**, 275 (1988).

109. S. Okubo, *Alg. Group. Geom.*, **1**, 62 (1984).

110. V. G. Kac, *Comm. Algebra*, **5**, 1375 (1977).

111. I. Shestakov, to appear in the *Proceedings of the 3rd International Conference on Non-associative Algebras and its Applications*, held at the University of Oviedo, Spain, August 1993.

Index

135